Untermassfeld

A late Early Pleistocene (Epivillafranchian) fossil site near Meiningen (Thuringia, Germany) and its position in the development of the European mammal fauna

Ralf-Dietrich Kahlke

Translated from German by Hans van Essen

Edited by Adrian Lister

BAR International Series 1578
2006

Published in 2016 by
BAR Publishing, Oxford

BAR International Series 1578

Untermassfeld

ISBN 978 1 84171 792 0

COVER IMAGE
Hippopotamus amphibius antiquus, *Untermassfeld. Right mandible with milk dentition (D_1-D_4) and erupting canine and first molar [IQW 1979/ 15 115 (Mei. 14 597)], buccal view. Scale: approx. 1:2*

Layout: D. Stein, Weimar

BAR Publishing is the trading name of British Archaeological Reports (Oxford) Ltd. British Archaeological Reports was first incorporated in 1974 to publish the BAR Series, International and British. In 1992 Hadrian Books Ltd became part of the BAR group. This volume was originally published by Archaeopress in conjunction with British Archaeological Reports (Oxford) Ltd / Hadrian Books Ltd, the Series principal publisher, in 2006. This present volume is published by BAR Publishing, 2016.

Printed in England

BAR
PUBLISHING

BAR titles are available from:

BAR Publishing
122 Banbury Rd, Oxford, OX2 7BP, UK
EMAIL info@barpublishing.com
PHONE +44 (0)1865 310431
FAX +44 (0)1865 316916
www.barpublishing.com

Contents

The excavation maps referred to on p. 110 are now available to download from: http://www.barpublishing.com/additional-downloads.html

Preface

The fossil site of Untermassfeld, near the town of Meiningen in Southern Thuringia, was discovered in 1978 and has been the subject of 25 field seasons, totalling nearly 100 months of excavation – a record that few of even the most celebrated of palaeontological or archaeological sites could match. The digs have produced a stunning array of fossil vertebrate remains in stratigraphic context, making this unquestionably one of the most important Quaternary localities in Europe. The project has been co-ordinated by Ralf-Dietrich Kahlke, Head of the Quaternary Palaeontology Research Station of the Senckenberg Institute, Weimar, and aided by a large international group of palaeontologists, biologists, geologists and geophysicists. Some of their detailed findings have appeared in three volumes of a site monograph, released in 1997 and 2001. In this volume, Kahlke provides the first full synthesis of the work, bringing the results up to date, and placing them in a broad context.

With some 14,000 determinable vertebrate fossils, the Untermassfeld assemblage represents the most complete assemblage of the time span 1.2-0.9 Ma BP in the Western Palaearctic. This dating is a key factor in understanding the importance of the site. First, on purely faunal grounds, the Untermassfeld assemblage lies squarely within the period of major turnover representing the transition from the Villafranchian to the Galerian – the single most important step in the origin of the modern fauna. However, as Kahlke clearly shows, many of the described taxa show evolutionary stages distinct both from those of latest Villafranchian and those of the early Middle Pleistocene. The Untermassfeld site therefore provides by far the richest source of evidence for a distinct faunal episode – the Epivillafranchian. An extremely valuable part of the work is a fully-referenced biostratigraphic comparison of the assemblage with those of 180 other localities of Villafranchian to Galerian age across Europe and western Asia.

The fossil collection, all of which is beautifully prepared and curated at the Weimar institute, provides important research and reference material for many animal groups. For many of the taxa substantial samples are available. Several new species of larger mammals have been described from this assemblage, including a large but gracile bison *Bison menneri,* and a species of large deer, *Eucladoceros giulii*. Other spectacular finds include the giant hyaena *Pachycrocuta*, represented not only by massive crushing jaws but also gnaw marks on prey bones and nests of coprolites. Big cats include sabre-tooths and forms akin to modern puma and cheetah.

The review also covers the geology and genesis of the site, detailed landscape reconstruction, and palaeoecological considerations. Particularly fascinating is the in-depth analysis of site taphonomy, with all finds (including many partially-articulated skeletons) plotted in three dimensions. It is a valuable exemplar of the role of fluviatile processes in accumulating fossil remains, with cadavers from upstream arriving at the site in various stages of decomposition. Kahlke draws not only on geological and palaeontological data, but also on modern examples of animal death and survival in flood conditions, to provide new insights into this still poorly-understood area of palaeontology.

Finally, and by no means least, the Untermassfeld site provides a benchmark for a fluvial locality shortly before the earliest known human occupation of western and central Europe. Excavated and documented to a level of detail equal to any archaeological locality, the site illustrates the nature of an accumulation by natural processes without any human impact. Bone distribution, taxon frequency, body-part representation, mortality spectra, surface damage and breakage: all of these will provide an essential point of reference for the interpretation of putative hominid sites of similar age.

In synthesising the vast repository of information on this cardinal locality, and its comparison to assemblages across the whole of Europe, Ralf-Dietrich Kahlke has done justice to his sterling efforts, and those of his collaborators, over 25 years of study, and in so doing has provided an invaluable document for Quaternary palaeontologists and archaeologists alike.

London, June 2006

Adrian Lister
Professor of Palaeobiology
University College London

Foreword and acknowledgements

According to our present knowledge, the time between 1.2 and 0.8 million years ago was characterized by the spread of early hominids into Europe. This process took place against a background of far-reaching structural changes within mammalian faunas. However, the understanding of the European animal world as it existed at the end of the Early Pleistocene and during the earliest Middle Pleistocene has remained vague for a long time, because larger palaeontological sites containing vertebrates were lacking.

The discovery of the Untermassfeld site near Meiningen in southern Thuringia (Central Germany) in 1978 triggered excavations that have lasted for more than 25 years and have yielded more than 14,000 fossil vertebrate remains in nearly 100 months of field work. Late Early Pleistocene fossil material is therefore now available in abundance, and allows detailed insights into the Epivillafranchian faunal history of the Western Palaearctic. The analysis of the finds from Untermassfeld was entrusted to an interdisciplinary study group whose results up to 1997 and 2001 were presented in three monographs.

The 18th International Senckenberg Conference on „Late Neogene and Quaternary biodiversity and evolution: Regional developments and interregional correlations“ was held at Weimar in 2004. During this conference the idea came up to make the results of current research on the faunal remains from Untermassfeld accessible to a wider circle of specialists in the form of a synopsis in the English language.

My thanks are due to all colleagues who partake in the Untermassfeld Study Group and whose results were incorporated in the present account, especially to G. Böhme (Berlin) (amphibians), B. Cramer (Tübingen) (determination of palaeotemperatures), I. A. Dubrovo (Moscow) (elephantids), J. Ellenberg (Jena) (sedimentology, regional geology), M. Faure (Lyon) (suids), N. García (Berkeley / Madrid) (ursids), C. Guérin (Lyon) (suids), H. Hemmer (Mainz) (felids), D. Jánossy † (Budapest) (birds), H.-D. Kahlke (Weimar) (cervids, rhinocerotids), J.-A. Keiler (Weimar) (preparation and conservation, coprolites), U. Kierdorf (Hildesheim) (palaeopathology), E. Krolopp (Budapest) (gastropods), L. C. Maul (Weimar) (turtles, micromammals, gnaw marks), R. Musil (Brno) (equids, ursids), E. Rutte (Würzburg) (fishes), A. V. Sher (Moscow) (bovids), M. A. Sotnikova (Moscow) (canids), E. Stephan (Constance) (determination of palaeotemperatures), A. Turner (Liverpool) (hyaenids), H.-P. Uerpmann (Tübingen) (determination of palaeotemperatures), F. Wiegank (Potsdam) (palaeomagnetism), M. Wolsan (Warsaw) (mustelids), and H. Zapfe † (Vienna) (cercopithecids).

For discussions, instructions and professional support on specific aspects, I am grateful to M. Böhme (Munich) (reptiles, fishes), V. Eisenmann (Paris) (equids), J. Ellenberg (Jena) (geomorphology), H. v. Essen (Dieren) (elephantids), N. García (Madrid) (carnivores, taphonomy, bibliographical support), S. Gaudzinski (Cologne / Neuwied) (mortality patterns), H. Hemmer (Mainz) (carnivores, ecology, ethology), H.-D. Kahlke (Weimar) (reconstructions of anatomically connected units), T. Krakhmalnaya (Kiev) (bibliographical support), J. v. d. Made (Madrid) (comparison of faunas), L. C. Maul (Weimar) (micromammals), S. Meng (Halle) (gastropods), D. Mol (Hoofddorp) (elephantids), O. Oms (Barcelona) (palaeomagnetism), A. Precker (Hamburg) (determination of potential inundation area), C. Radulescu † (Bucharest) (regional faunas), M. V. Sotnikova (Moscow) (comparison of faunas), M. Stebich (Weimar) (botany), and A. K. Vekua (Tbilisi) (regional faunas).

As employees of the Senckenberg Research Unit for Quaternary Palaeontology, Weimar (until 1990 the Institute for Quaternary Palaeontology, Weimar, and from 1990-1999 the Department for Quaternary Palaeontology, Weimar, of the Friedrich Schiller University in Jena) the staff members mentioned in footnote 1 (Section 4.1.) have been involved in drawing up the excavation plans for more than 20 years. They deserve special thanks. For the preparation of thousands of bones I am indebted to M. Brohm, V. Karl, J.-A. Keiler, V. Nee, K. Neumann, H. Nolte †, U. Rohmer, D. Rössler, M. Stache, G. v. Stuckrad, G. Utschig, and H.-J. Wolfram † (all Weimar). The diagrams on the osteological material were drawn by the skilled hand of H. Wöllner, the final drawings for the figures 49-72 and 77 as well as the appended excavation plans mostly by N. Fiebig (both Weimar). For taking the large majority of the photos I thank T. Korn (Weimar). G. Utschig (Weimar) was so kind as to take care of the picture editing. Additional digital support was given by M. Stache (Weimar).

Special thanks are due to H. v. Essen (Dieren), who with accustomed perfection translated the German text into English. I also extend my gratitude to A. Lister, editor of the present volume, who initiated its publication in Great Britain. We are indebted to BAR, Oxford for their willingness to accept this contribution for the International Series of the British Archaeological Reports. I very cordially thank D. Stein (Weimar) for producing the layout.

For financing the excavations near Untermassfeld I thank the Senckenberg Research Institute and Natural History Museum, the Free State of Thuringia (Germany), and the Town of Meiningen. The last of these, represented by its treasurer K.-D. Schmidt, is also acknowledged for financially supporting the printing of the present monograph.

Weimar, June 2006

Ralf-Dietrich Kahlke

1. Introduction

In January 1978 an exceptionally rich vertebrate site of Early Pleistocene age was discovered 2 km south of the town of Meiningen (southern Thuringia), northeast of the village Untermassfeld on the right slope of the valley of the river Werra. After initial salvage excavations, the site has been laid open and documented continuously in yearly digging campaigns of mostly five months each. From 1983 these were conducted by the Institute for Quaternary Palaeontology, Weimar, and from 1990 by the Department for Quaternary Palaeontology Weimar, coming under the Institute for Geosciences of the Friedrich Schiller University in Jena. From the year 2000 the excavations have been continued by the Research Unit for Quaternary Palaeontology, Weimar, as part of the Senckenberg Research Institute (R.-D. Kahlke 1997b, 2001c). In the autumn of 1995 the excavated area measured 584 m^2, and depths of maximally 5.0 m below surface level were reached. Since that time the area of the excavation has not been enlarged, but its depth has been increased to 5 m below the surface (Fig. 1). Up to the said date, well over 9,400 identifiable remains of large mammals, as well as approximately 3,000 dental and skeletal remains of small mammals and other small vertebrates, had been recovered. Preparation and conservation of the material have been carried out at the research institute in Weimar (Keiler 1995, 2001a, 2001b). Since 1994 the complex site of Untermassfeld, including its immediate surroundings, has been registered as a protected geo-monument of the Free State of Thuringia.

The diversity, quality, and quantity of the recovered fossil material, as well as the general rarity of larger find complexes of later Early Pleistocene age in Eurasia, and especially in Central Europe, required, if possible, a highly complex interdisciplinary analysis of the finds and results. Already in 1980 an international Untermassfeld Research Group was established, and extended during subsequent years (coordinator: H.-D. Kahlke, from 1991 onward R.-D. Kahlke). Work in the following disciplines was allotted as separate projects (situation at 30 March 2006):

Fig. 1 Untermassfeld.
Part of the excavation area, approx. 4.5 m below surface level (squares Q 21-24, Q 47-50, Q 155-158, September 2002). Photography by courtesy of M. Steinmetz, Hamburg.

1. Excavation and documentation	(Teams of Senckenberg Weimar)	*		***
2. Regional and local research history	(R.-D. Kahlke, Weimar)	*	**	
3. Regional geology	(J. Ellenberg, Jena)	*		
4. Geology of the site	(R.-D. Kahlke, Weimar)	*	**	
5. Sedimentology	(J. Ellenberg, Jena)	*		
6. Palaeomagnetics	(F. Wiegank, Potsdam)	*		
7. Absolute age determination	(C. Falguères, Paris & J.-J. Bahain, Paris)			→
8. Reconstructions of vegetation	(M. Stebich, Weimar & R.-D. Kahlke, Weimar)			→
9. Ostracods	(E. Pietrzeniuk, Berlin)	↔		
10. Gastropods	(E. Krolopp, Budapest)	*		
11. Fishes	(E. Rutte, Würzburg)	*		
12. Amphibians	(G. Böhme, Berlin)	*		
	(M. Böhme, Munich)			→
13. Lizards and snakes	(M. Böhme, Munich)			→
14. Testudinates	(L. C. Maul, Weimar)	*		
15. Birds	(D. Jánossy, Budapest)	*		
	(J. Mlikovsky, Prague)			→
16. Bovids	(A. V. Sher, Moscow)	*		→
17. Cervids	(H.-D. Kahlke, Weimar)	*	**	a)
18. Hippopotamids	(R.-D. Kahlke, Weimar)	*	**	
19. Suids	(C. Guérin, Lyon & M. Faure, Lyon)	*		
20. Rhinocerotids	(H.-D. Kahlke, Weimar)	*		
21. Equids	(R. Musil, Brno)		**	
	(V. Eisenmann, Paris)			→
22. Elephantids	(I. A. Dubrovo, Moscow)		**	
	(H. v. Essen, Dieren, D. Mol, Rotterdam, A. Lister, London & E. N. Maschenko, Moscow)			→
23. Canids	(M. V. Sotnikova, Moscow)		**	→
24. Ursids	(R. Musil, Brno)		**	
	(N. García, Madrid)		b)	→
25. Mustelids	(M. Wolsan, Warsaw)		**	
26. Hyaenids	(A. Turner, Liverpool)		**	
27. Coprolites	(J.-A. Keiler, Weimar)		**	
28. Felids	(H. Hemmer, Mainz)		c)	***
29. Insectivores, lagomorphs, and rodents	(L. C. Maul, Weimar)			***
30. Primates	(H. Zapfe †, Vienna)			***

31. Taphonomy	(R.-D. Kahlke, Weimar & S. Gaudzinski, Neuwied)		d)	***
32. Tooth root cementum analysis	(H. Kierdorf, Cologne) (N. García, Madrid)	↔		 ↔
33. *Stephanorhinus* mesowear	(T. Kaiser, Greifswald)			↔
34. Osteophagia by herbivores	(R.-D. Kahlke, Weimar)			***
35. Rodent gnaw marks	(L. C. Maul, Weimar)			***
36. Palaeopathology	(U. Kierdorf, Hildesheim)		e)	→
37. Palaeotemperature determination	(E. Stephan, Constance, H.-P. Uerpmann, Tübingen & B. Cramer, Tübingen)			***
38. Palaeoecology	(R.-D. Kahlke, Weimar)	*		***
39. Stratigraphy	(R.-D. Kahlke, Weimar)	*		***
40. Methods of preparation and conservation	(J.-A. Keiler, Weimar)		f)	***

Legend:

*	results published in part 1 of the Untermassfeld monograph [R.-D. Kahlke (ed.) 1997a]
**	results published in part 2 of the Untermassfeld monograph [R.-D. Kahlke (ed.) 2001a]
***	results published in part 3 of the Untermassfeld monograph [R.-D. Kahlke (ed.) 2001b]
→	in progress
a)	H.-D. Kahlke 1995a
b)	Garcia 2004
c)	Hemmer et al. 2003
d)	R.-D. Kahlke 1999; R.-D. Kahlke & Gaudzinski 2005
e)	R.-D. Kahlke & Kierdorf 1997, Kierdorf & R.-D. Kahlke 2004
f)	Keiler 1995, 2001a
↔	Available material did not lead to results

Interim results and first contemplations, with a wider scope than the separate projects, were presented during the Fifth International Palaeontological Colloquium entitled „The Pleistocene of Untermassfeld near Meiningen – On the biostratigraphical correlation of Early Pleistocene sites within Eurasia", held from 1 to 7 September 1985 in Weimar, and additionally in the form of more talks and first publications as well as various specialist excursions and special exhibitions (detailed account in R.-D. Kahlke 1997b, 2001c, 2002a). Selected contributions to the Fifth Palaeontological Colloquium – concerning Eurasian Early Pleistocene vertebrate sites or animal groups represented in the Untermassfeld fauna – were published as a thematic volume within the series "Quartärpaläontologie" [H.-D. Kahlke (ed.) 1990a].

In a way that has been unique until now, Untermassfeld documents a critical phase in evolutionary history, during which European faunas of the latest Villafranchian were modified into the more modern early Middle Pleistocene ones. The site is characterized by a number of aspects that make further excavations appear advisable. For example, to judge by the distributional pattern so far observed within the fossil material (cf. Section 4.3.), one may count on the presence of highly informative and concomitantly well-preserved skeletal remains in the remaining parts of the fluviatile sequence. Finds that could complement the material that has been recovered to date are especially desirable for the following groups: skulls and sets of teeth of adult hippopotamids, rhinocerotids, equids, and elephantids; complete antlers of all cervid species; remains of roe deer and elks in general; postcranial skeletal elements of some felids; and evidence of small carnivores as well as primate remains. The so far largely uncomplicated accessibility of the fossiliferous sands, their great stability as well as the usually exquisite state of preservation of the fossils through impregnation with carbonates, may be described as ideal.

Now that the essential initial studies of the complex site of Untermassfeld have been completed, the obvious next goals are an overview of the achieved results, as well as a synthetic approach towards the genesis of the fossil lagerstätte, ecological parameters including the landscape characteristics, and the stratigraphical position of the recovered fauna. These are the goals of the present volume.

2. Geology and palaeomagnetism

In terms of regional geology, the middle region of the Werra valley with the site of Untermassfeld belongs to the southwestern Thuringian Triassic terrain, which forms part of the South German fault block (Kästner et al. 1995). Apart from faults of Saxonian age that show Hercynian (NW-SE) strikes and branch off the Franconian Line (the southwestern marginal fault of the Thuringian Forest and the Thuringian-Franconian Slate Mountains), there are faults in southern Thuringia that show NNE-SSW or NNW-SSE strike directions. Hence it cannot be ruled out that the change of direction in the course of the Werra near Untermassfeld (from E-W to S-N; cf. Fig. 78) is an after-effect related to a pre-Quaternary fault pattern.

At Untermassfeld the eastern valley slope of the Werra cuts Triassic formations. Below the site, Middle Triassic calcareous rocks (Lower Muschelkalk: Lower and Middle Wellenkalk with Oolitic Zone = mu_1, mu_1oo, mu_2) outcrop above horizons of the Lower Triassic (Upper Bunter sandstone: *Myophoria* Sequence). Above the site the entire remaining sequence of the Lower Muschelkalk (mu_2-$mu_3\chi$), up to the *orbicularis* Strata (mm), follows in normal arrangement.

The lithostratigraphical subdivision of the often subrosively changed Quaternary in the Werra valley is mainly based on geomorphological, Quaternary-geological and sedimentological analyses of the forms and outcrops present between Schwallungen and Neustädt (Ellenberg 1968). Directly below the level of the excavation, the base of the Pleistocene profile is formed by 7-8 m thick deposits of weathered coarse gravels (the so-called "Jüngere Zersatzgrobschotter" = "Yonger Weathered Coarse Gravels") at about 359 m above O.D., i.e., at a relative height of 69 m above the present floodplain (Fig. 2). According to the outlined succession of terraces in the middle Werra valley (Ellenberg & R.-D. Kahlke 1997, 31ff.), deposition took place during the Eburonian. Deep chemical weathering with a destructive impact on fossils supervened during the predominantly warm Waalian (cf. also Section 7.1.).

The coarse gravel deposits are overlain by a macroscopically so far unfossiliferous clayey to silty floodplain sediment which is of uncertain stratigraphical position and may reach a thickness of 2 m in close vicinity to the site. The fluviatile sands cut by the excavation are normally contiguous.

In the sand pit / refuse dump directly north of the excavation area, as well as at the site itself, a 2.5-8.0 m thick multicoloured series of alternating fine- to medium-grained sands constitute the main part of the fluviatile deposits. The lithological characteristics of these so-called Lower Fluviatile Sands demonstrate that they are floodplain deposits (Ellenberg & R.-D. Kahlke 1997, 37ff., 57). The flow rates of the sediment-depositing water may be specified as 0.4-0.5 m/s. Both the variable thickness of the individual layers, and an increase in average grain size proportional to the thickness of the separate layers, demonstrate flooding at higher or lower energy levels. In various places, comparatively small festoon cross-bedded channel infills cut through the sands. From several sand layers below and above a widely demonstrable silt horizon, generally very well-preserved bones have been recovered.

The main bone bed is constituted by Upper Fluviatile Sands that form an infill of an erosional channel which cut into the Lower Fluviatile Sands to a depth of 4.0 m, and runs more or less parallel to the present course of the river. In the northwestern area of the site, and below the level of the main excavation plane, the erosional channel cut through the underlying floodplain sediment and into the uppermost decimetres of the weathered coarse gravels (R.-D. Kahlke 2001d). Here a sandy-gravelly amalgamation zone full of bone remains was formed in the deepest part of the channel. The lower part of the Upper Fluviatile Sands, up to 3.0 m thick, is formed by smoothly and horizontally layered reddish brown medium-grained sand with an admixture of fine-grained sand. Granulometric analysis demonstrates that the environment in which sedimentation took place was fluviatile and characterized by decreasing water depth and increasing distance from the midstream, i.e., close to the riverbank. The highest concentrations of fossil skeletal elements (up to > 100 finds/m^2) were generally recorded in the basal part of this Channel Infill.

A layer of brown silty sand, on average 0.2 m thick, sets off the upper part of the Upper Fluviatile Sands, which is only present in the western to southwestern area of the excavation (i.e., on the side that faces the valley). Because it was impermeable to water, the so-called "Silt Band" stimulated soil movements in the overlying sediment, but at the same time protected the directly underlying lower part of the Upper Sands, as well as the fossils contained in it, against solifluction. Within the upper section of the Upper Fluviatile Sands, the multicoloured medium-grained sands reach a thickness of up to 2.0 m and until now have proved to be non-fossiliferous. Their primary bedding structure was annihilated by solifluction that turned them into a series of sinuous sediment bodies.

In the southern part of the excavation area the highly fossiliferous lower section of the Upper

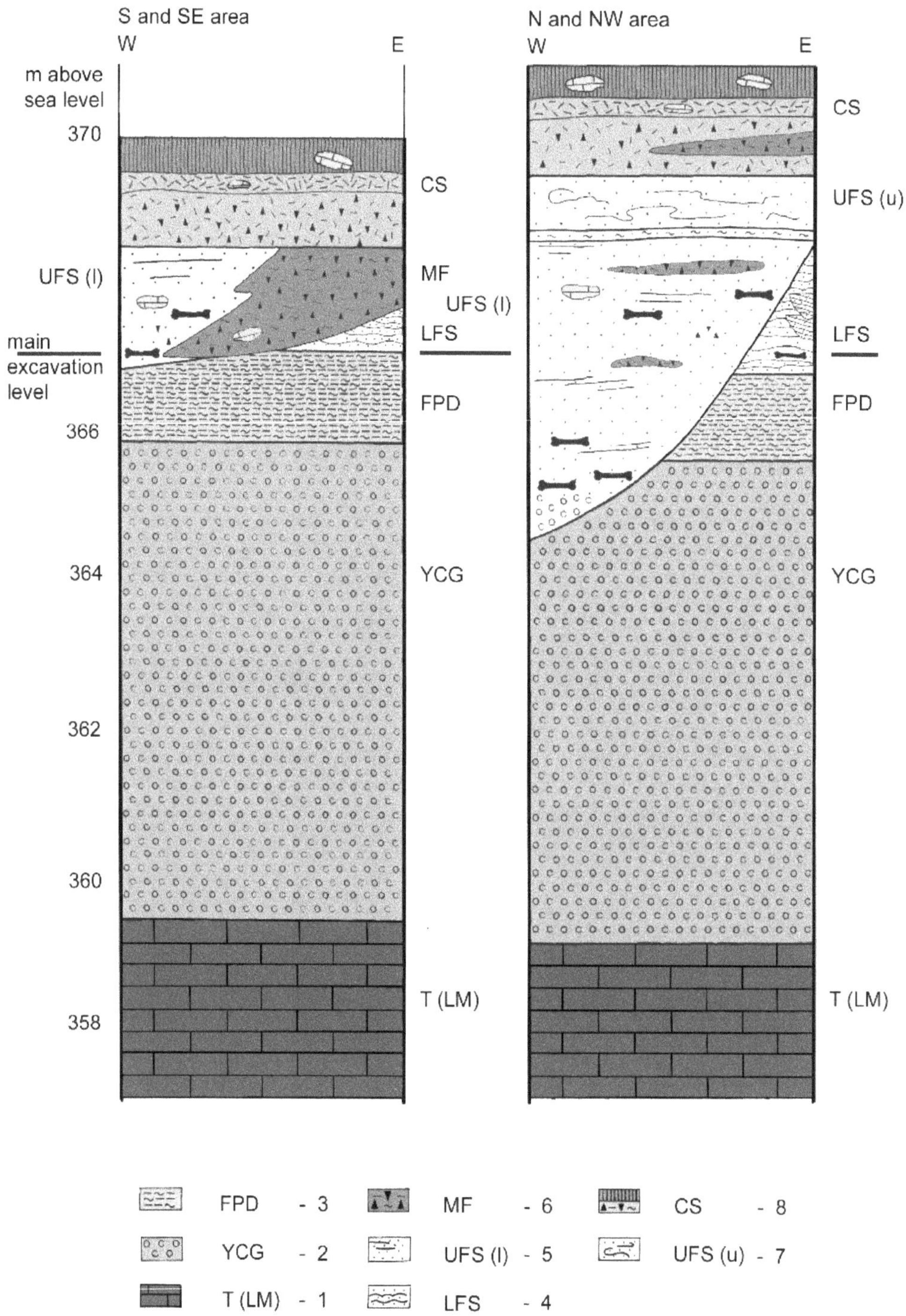

Fig. 2 Untermassfeld.
Diagrammatic sections of the excavated area (neotectonic movements eliminated).
Legend: 1 T (LM): Limestone (Middle Triassic, Lower Muschelkalk). – 2 YCG: Younger Coarse Gravels. – 3 FPD: Floodplain deposits. – 4 LFS: Lower Fluviatile Sands. – 5 UFS (l): Upper Fluviatile Sands (lower part). – 6 MF: Coarse Clastic Mudflow Fan. – 7 UFS (u): Upper Fluviatile Sands (upper part). – 8 CS: Series of cover sediments.

Fluviatile Sands interfingers with a Coarse Clastic Mudflow Fan ("Hangschuttfächer"), which has been exposed to a depth of 1.0 m. This sediment body consists of calcareous and clayey to silty erosion products from the Muschelkalk slope. As an accumulating factor it influenced the hydrodynamical conditions at the time the fluviatile series was deposited in the vicinity of the present site (cf. Section 5.1.). In the leeward (northern) margin of the fan, the zone of bone accumulations commences. Overall characteristics of the fossiliferous section of the Upper Fluviatile Sands are wedge-shaped layers of Muschelkalk debris as well as the regular occurrence of single larger boulders from the marker strata that outcrop above the site (*Terebratula* zone = mu_2t, Schaumkalk zone

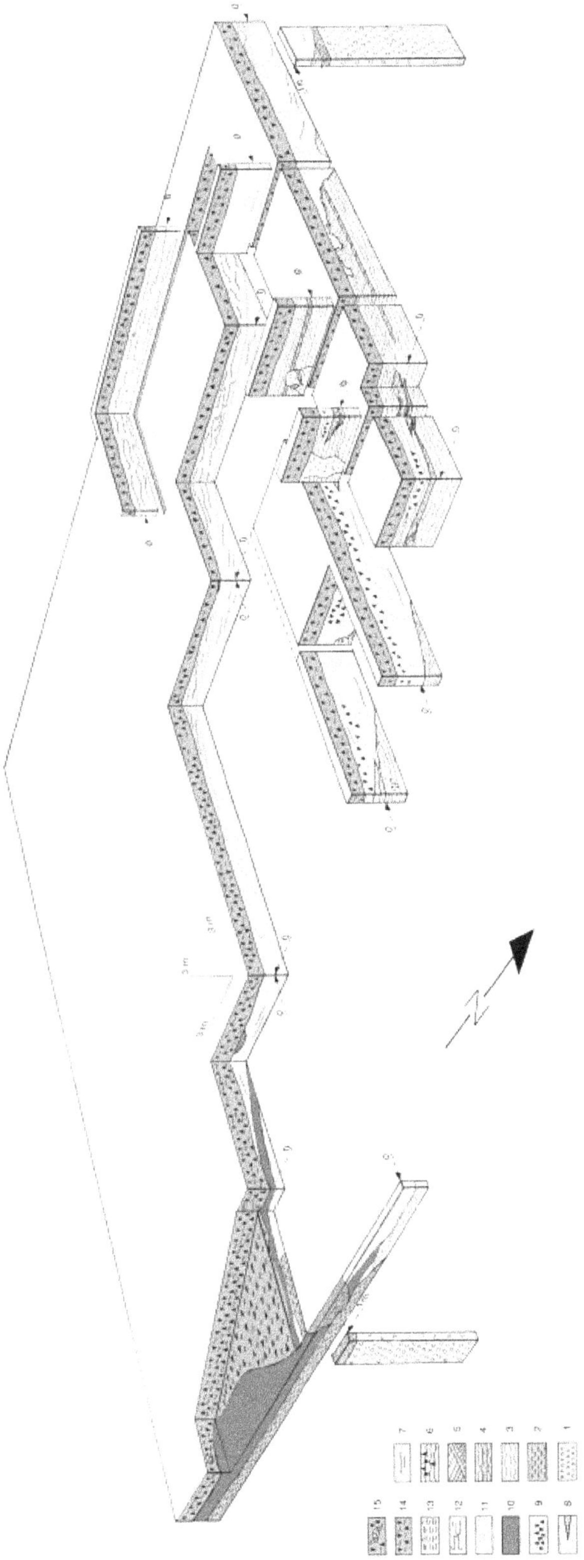

Fig. 3 Untermassfeld.
Simplified spatial representation of the strata exposed by the excavation (situation at the end of 1995).
Legend: 1 Younger Coarse Gravels (YCG). – 2 Floodplain deposits (FPD). – 3 Lower Fluviatile Sands (LFS). – 4 Sandy silt in Lower Fluviatile Sands (LFS). – 5 Greyish brown, fine- to medium-grained, crossbedded sand (LFS). – 6 Limestone (Muschelkalk) debris in Lower Fluviatile Sands (LFS). – 7 Upper Fluviatile Sands, lower part [UFS (l)]. – 8 Reddish yellow, medium-grained sand layer in Upper Fluviatile Sands (UFS). – 9 Limestone (Muschelkalk) debris in Upper Fluviatile Sands (UFS). – 10 Coarse Clastic Mudflow Fan (MF). – 11 „Silt band“ (UFS). – 12 Upper Fluviatile Sands, upper part [UFS (u)]. – 13 Silty to clayey channel infill. – 14 Series of cover sediments (CS). – 15 Boulders in cover sediments (CS).

= $mu_3\chi$). The material could have been eroded from the Clastic Mudflow Fan or have entered the deposit directly from the valley slope.

Wherever they outcrop, the Fluviatile Sands, as well as the Clastic Mudflow Fan that wedges into them, are covered by a 1.0-2.0 m thick deposit of talus marl which includes layers of Muschelkalk debris that thin out towards the valley.

In the sand pit / refuse dump north of the excavation area, the Quaternary sediments slightly dip towards the southwest as a result of neotectonic movements in the foreland of the Marisfeld fault zone (Fig. 3). Within the sequence of strata cut by the excavation (floodplain sediment to Upper Fluviatile Sands), the southwestward dip clearly increases from NE to SW (maximum value 130/20 SW). Throws that may sometimes reach heights of 0.4 m bridge zones of intensified stress. Plastic sediment deformations that also produced effects on larger bone accumulations or larger skeletal elements (cf. Section 4.2.) were particularly observed at the northeastern channel border, which faces the slope. As a result of subrosion, i.e. erosion of readily soluble sediments in the subsoil, bowl-shaped depressions of a few metres in diameter, or smaller funnels caused by the caving in of sediments, occasionally occur in the Lower Fluviatile Sands of both the sand pit and the site. It is very likely that corresponding precursory cavities were for the large part brought about by material leaching away from the Lower Triassic (Röt) salt deposit.

85 samples of unconsolidated sediment taken from the Younger Weathered Coarse Gravels [outcrop of the so-called Middle Clay Prominence ("Mittlerer Tonkopf"), 0.8 km north of the excavation] as well as from the Lower and Upper Fluviatile Sands (sand pit, excavation) were analysed with regard to their magnetization. According to Wiegank (1997) the following overall diagnosis of the palaeomagnetic situation emerges:

Both the Younger Weathered Coarse Gravels (which underlies the sandy sequence) and the basal and middle parts of the Lower Fluviatile Sands show reversed magnetization. On the other hand, the fossiliferous upper part of the Lower Fluviatile Sands (high flood sediments) as well as the highly fossiliferous lower section of the Upper Fluviatile Sands (Channel Infill) are normally magnetized (see also the diagram in R.-D. Kahlke 1995a, fig. 13; 1999, fig. 1).

3. Palaeontology

3.1. Flora

All Early Pleistocene layer sections cut by the Untermassfeld excavation were sampled, but none yielded a sufficient quantity of analysable pollen (M. Stebich, Weimar). Neither could any statistically relevant pollen spectra be obtained from the recovered hyaena coprolites (see below). Moreover, there are no determinable macroscopic remains of fossil flora from Untermassfeld. Only in the interior and on the surface of some coprolites are plant imprints occasionally found. In one such case a leaf blade fragment of a grass was identified (Keiler 2001c, pl. 131, 1-2).

In the bone-bearing upper section of the Lower Fluviatile Sands, specks or lenticular concentrations of carbonaceous particles (xylite) were occasionally observed. These embedded seams are maximally 10 mm thick and cover areas of a few dm^2. It is very likely that they represent plant remains (leaves, needles, bark, twigs, etc.) that were washed out to certain spots and then covered by sediment within a relatively short time. In the Channel Infill, xylite particles with sizes of some mm regularly occur, and their number markedly increases within the bone concentrations. In this case, too, it is likely that remains of a former riverine vegetation (bark, branches, shrubs, etc.) were preserved.

In addition, etched traces on the surfaces of several bones (cf. R.-D. Kahlke 1997c, pl. 61, 9) were caused by root systems of plants and demonstrate the existence of vegetation after the finds had been covered by sediment. Root hairs also penetrated coprolites embedded in the sediment, as shown by electron microscopic scans of matching tubular cavities (Keiler 2001c, pl. 131, 3-4).

3.2. Fauna

According to present knowledge, the range of the fossil faunal remains from Untermassfeld comprises the following taxa (situation per 30 March 2006):

Class: G a s t r o p o d a

Order: Mesogastropoda

Valvata cristata O. F. Müller, 1774
Bithynia leachi (Sheppard, 1823)
Belgrandia germanica (Clessin, 1882)
? *Pomatias* sp.
Acicula polita (Hartmann, 1870)

Order: Basommatophora

Lymnaea peregra (O. F. Müller, 1774)
Lymnaea truncatula (O. F. Mülller, 1774)
Planorbis planorbis (Linnaeus, 1758)
Anisus leucostoma (Millet, 1813)
Bathyomphalus contortus (Linnaeus, 1758)
Carychium minimum O. F. Müller, 1774

Order: Stylommatophora

Succinea oblonga Draparnaud, 1801
Catinella arenaria (Bouchard-Chantereaux, 1837)
Cochlicopa lubricella (Porro, 1837)
Granaria cf. *frumentum* (Draparnaud, 1801)
Pupilla muscorum (Linnaeus, 1758)
Vallonia pulchella (O. F. Müller, 1774)
Vallonia costata (O. F. Müller, 1774)
Clausilia dubia Draparnaud, 1805
Clausiliidae indet.
Cecilioides acicula (O. F. Müller, 1774)
Discus ruderatus (Férussac, 1821)
Nesovitrea hammonis (Ström, 1756)
Zonitidae indet.
Limacidae indet.
Euconulus fulvus (O. F. Müller, 1774)
Bradybaena cf. *fruticum* (O. F. Müller, 1774)
Helicella sp. indet.
Helicopsis sp. indet.
Trichia hispida (Linnaeus, 1758)
Trichia sp. indet.
Helicigona lapicida (Linnaeus, 1758)
Cepaea cf. *nemoralis* (Linnaeus, 1758)
Helicidae indet. (1)
Helicidae indet. (2)
Helicidae indet. (3)

Class: O s t e i c h t h y e s

Order: Esociformes

Esox lucius Linnaeus, 1758

Order: Cypriniformes

aff. *Gobio* sp.
Scardinius sp.

Class: A m p h i b i a

Order: Urodela

Triturus cf. *cristatus* (Laurenti, 1768)
Triturus cf. *vulgaris* (Linnaeus, 1758)
one more species (M. Böhme, in progress)

Order: Anura

Bufo cf. *bufo* (Linnaeus, 1758)
Bufo cf. *viridis* Laurenti, 1768

Hyla cf. *arborea* (Linnaeus, 1758)
Rana cf. *arvalis* Nilsson, 1842
Rana cf. *dalmatina* Bonaparte, 1840

Class: Reptilia
Order: Squamata
Chalcides sp.
Eublepharis sp.
three more species (M. Böhme, in progress)

Order: Testudinata
Emydini gen. et sp. indet.

Class: Aves
Order: Anseriformes
Cygnus sp. (cf. *olor* Gmelin, 1788)
Anser sp. (aff. *subanser* Jánossy, 1983)

Order: Accipitriformes
Haliaeëtus sp. (aff. *brevipes* Jánossy, 1983)

Order: Galliformes
Francolinus (Lambrechtia) capeki (Lambrecht, 1933)

Order: Passeriformes
Turdus sp. (aff. *philomelos* Brehm, 1831)
Corvus sp. (aff. *janossyi* Mourer-Chauviré, 1975)
Garrulus aff. *glandarius* (Linnaeus, 1758)
one more species (J. Mlikovsky, in progress)

Class: Mammalia
Order: Artiodactyla
Bison menneri Sher, 1997
Capreolus cusanoides Kahlke, 2001
Alces carnutorum (Laugel, 1862)
Cervus s.l. *nestii vallonnetensis* de Lumley, Kahlke, Moigne et Moullé, 1988
Eucladoceros giulii Kahlke, 1997
Hippopotamus amphibius antiquus Desmarest, 1822
Sus scrofa priscus Goldfuss, 1823

Order: Perissodactyla
Stephanorhinus etruscus (Falconer, 1859) (sensu H.-D. Kahlke 2001b)
Equus wuesti Musil, 2001
Equus sp.

Order: Proboscidea
Mammuthus meridionalis (Nesti, 1825) ssp.

Order: Carnivora
Panthera onca gombaszoegensis (Kretzoi, 1938)
Acinonyx pardinensis pleistocaenicus (Zdansky, 1925)
Puma pardoides (Owen, 1846)
Lynx issiodorensis (Croizet et Jobert, 1828) ssp. ex aff. *spelaeus* (Boule, 1906)
Megantereon cultridens adroveri Pons Moyá, 1987
Homotherium crenatidens (Weithofer, 1898) ssp.
Pachycrocuta brevirostris (Aymard, 1846)
Meles hollitzeri Rabeder, 1976
Ursus cf. *dolinensis* García et Arsuaga, 2001 (= *U. rodei* Musil, 2001)
Canis (Xenocyon) lycaonoides (Kretzoi, 1938)
Canis mosbachensis Soergel, 1925

Order: Rodentia
Sciurus cf. *whitei* Hinton, 1914
Spermophilus (Urocitellus) ex gr. *primigenius* (Kormos, 1934) / *polonicus* (Gromov, 1965)
Castor fiber Linnaeus, 1758
Trogontherium cuvieri Fischer von Waldheim, 1809
Hystrix sp.
Glis sackdillingensis (Heller, 1930)
Cricetus cf. *runtonensis* Newton, 1909
Clethrionomys cf. *hintonianus* Kretzoi, 1958
Pliomys episcopalis Méhely, 1914
Mimomys savini Hinton, 1910
Mimomys (Cseria) pusillus (Méhely, 1914)
Microtus thenii Malez et Rabeder, 1984
Apodemus cf. *sylvaticus* (Linnaeus, 1758)
one more species (L. C. Maul, in progress)

Order: Lagomorpha
cf. *Lepus* sp.

Order: Primates
Macaca sylvanus (Linnaeus, 1758) ssp. indet.

Order: Insectivora
Erinaceus cf. *europaeus* Linnaeus, 1758
Talpa cf. *europaea* Linnaeus, 1758
Talpa cf. *minor* (Freudenberg, 1914)
Sorex runtonensis Hinton, 1911
Sorex (Drepanosorex) ex gr. *margaritodon* Kormos, 1930 – *savini* Hinton, 1911 (transitional form)
Beremendia cf. *fissidens* (Petényi, 1864)

3.2.1. Gastropods

The abundance of gastropod remains in the fluviatile sequence of Untermassfeld is only moderate. In view of the high sedimentation rate, especially at the time the erosional channel was filled in, the large majority of the shell remains probably came into the area of the present site post mortem. At any rate, with 36 taxa (Krolopp 1997) the analysed fossil assemblage offers a fairly wide, but with respect to the Early Pleistocene valley of the Werra by no means comprehensive range. The eight aquatic species demonstrated represent only 4.9 % of the total number of individuals. Twenty-eight terrestrial forms, whose shell preservation precludes transport over great distances, indicate that materials came in from the area nearby. Small amber snail (*Succinea oblonga*) and blind awlsnail (*Cecilioides acicula*) predominate.

3.2.2. Fishes

With pike and carp (Esociformes, Cypriniformes) the ichthyofauna (Rutte 1997) includes only a minor number of the species that may be postulated for the Early Pleistocene Werra. The pike (*Esox lucius*) predominates with 33 identifiable remains of dentition and skeleton. According to M. Böhme (Munich) (in litt. 16. 10. 1998), *Tinca* sp. is not represented among the finds described by Rutte (l.c.). An argument against the attribution of a pharyngeal tooth [IQW 1987/ 21 979 (Mei. 21 498)] to *Leuciscus* (Rutte 1997, 79) is its robust "cutting edge", which is discernable in occlusal view and turned towards the medial side. Because among the recent indigenous carp a similar cutting edge is formed only on the hindmost pharyngeal teeth (pt 1 and pt 2) of the gudgeon, M. Böhme (l.c.) refers the specimen to aff. *Gobio* sp. Distinct indications of resorption on the said find strongly suggest natural tooth replacement, so that the fish is likely to have belonged to the autochthonous stock of the river (M. Böhme l.c.). In addition, G. Böhme (1997b, 122ff.) mentions a rudd-like species (*Scardinius* sp.) from Untermassfeld.

3.2.3. Amphibians

G. Böhme (1997a) had 151 isolated remains of amphibians at his disposal. With some reservation, evidence of the presence of a few salamanders (Urodela) may be considered to relate to the great crested newt (*Triturus* cf. *cristatus*) or otherwise the smooth newt (*Triturus* cf. *vulgaris*). With 130 identifiable finds the frogs (Anura) are predominant. They are represented by the common toad (*Bufo* cf. *bufo*), the green toad (*Bufo* cf. *viridis*), the tree frog (*Hyla* cf. *arborea*), the moor frog (*Rana* cf. *arvalis*), and the agile frog (*Rana* cf. *dalmatina*).

3.2.4. Reptiles

At present only a single hyoid fragment of a freshwater turtle (Emydini) demonstrates the occurrence of a reptile (Maul 1997). Other finds, recovered during the nineteen-eighties, are to be considered as lost after they had been given in loan. Among other taxa a skink (*Chalcides* sp.) and a gecko (*Eublepharis* sp.) were identified as exotic species (cf. the lecture account in Sher 1986, 139-140). Sifting through many small vertebrate remains newly recovered from the site produced evidence of three additional species (M. Böhme, in progress).

3.2.5. Birds

Consistent with the taphonomy of a fluviatile fossil lagerstätte, the relatively sporadic bird remains from Untermassfeld reflect only a small, though not atypical part of the expected avifauna. Of the approximately 60 available remnants, 40 specimens in total are to be attributed to waterfowl and screamers (Anseriformes), hawks (Accipitriformes), fowl-like birds (Galliformes), and larger sparrows (Passeriformes) (Jánossy 1997). Single specimens establish the presence of (mute) swan [*Cygnus* sp. (cf. *olor*)], a large goose species [*Anser* sp. (aff. *subanser*)], and the sea eagle [*Haliaeëtus* sp. (aff. *brevipes*)]. Of special stratigraphical and biogeographical importance is the so-far westernmost European find of an Early Pleistocene francolin [*Francolinus (Lambrechtia) capeki*]. The thrush [*Turdus* sp. (aff. *philomelos*)] is also represented, again by only one bone fragment. On the other hand, several partly associated skeletal elements were recovered that belong to a crow between carrion crow / rook and raven [*Corvus* sp. (aff. *janossyi*)] in size. In addition, two finds demonstrate the occurrence of the (Eurasian) jay (*Garrulus* aff. *glandarius*).

3.2.6. Mammals

Special attention has been paid to the mammalian fauna from Untermassfeld. For the study of the b o v i d s more than 1000 finds were available, and all of these are referable to a single species of *Bison* (Sher 1997). During the digging campaigns of 2002 and 2004 two skulls of adult animals were recovered (Fig. 4) (R.-D. Kahlke 2002b, 2004b; Sher, in progress). The morphologically homogeneous material shows a marked sexual dimorphism (Fig. 5). Sher (1997) conferred the new species name *Bison menneri* upon this form, which is large and long-legged, but rather gracile in comparison with statigraphically younger representatives of the genus. According to present knowledge, *Bison menneri* was the earliest representative of true bison (subgenus *Bison*) in Eurasia. Some older, smaller, and lighter forms from other Eurasian sites (Siwalik series, Nihewan, Taman Peninsula, Pirro Nord, etc.) are referable to the *Eobison* group (subgenus or genus) (Sher 1997, 175).

Material that belongs to c e r v i d s is very numerous as well. To date, the presence of four species has been established on the basis of a very variable amount of finds per species (H.-D. Kahlke 1997, 2001a). To begin with, an antler beam attached to the skull (Fig. 6), upper and lower tooth rows, as well as a few postcranial elements, represent an early form of roe deer, which was described as a new species (*Capreolus cusanoides*) by H.-D. Kahlke (2001a). Morphologically and metrically this form is intermediate between the Ruscinian / Early Villafranchian *Procapreolus cusanus* and the early Middle Pleistocene *Capreolus suessenbornensis*.

The body size of the likewise only sparsely documented elks touches the lower limit of the variation range of *Alces latifrons* from the early Middle Pleistocene. However, the form from Untermassfeld clearly shows a more gracile skeletal build (Fig. 7) (H.-D. Kahlke 1997, 184-195). During the 1995 digging campaign the left mandible ramus from Untermassfeld published by H.-D. Kahlke (1995a) was followed up by another left mandible fragment with P_3-M_3 of slightly larger dimensions (H.-D. Kahlke 2001a). Under the name of *Alces carnutorum* these finds therefore document a still little-known evolutionary stage of the elks, which is intermediate between the Villafranchian *Alces gallicus* and the younger *A. latifrons* (cf. also Breda & Marchetti 2005).

A smaller cervid is fairly often represented at

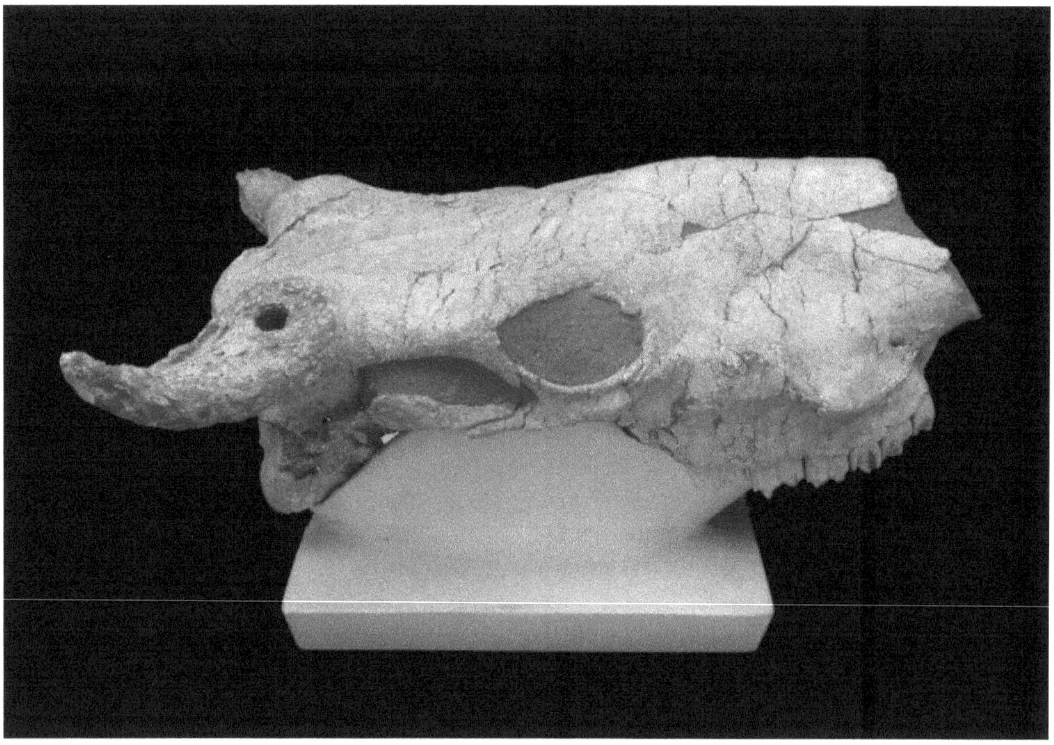

Fig. 4 *Bison menneri*, Untermassfeld.
Skull of an adult individual [IQW 2003/ 28 250 (Mei. 27 412)]. Scale: approx. 1:4.

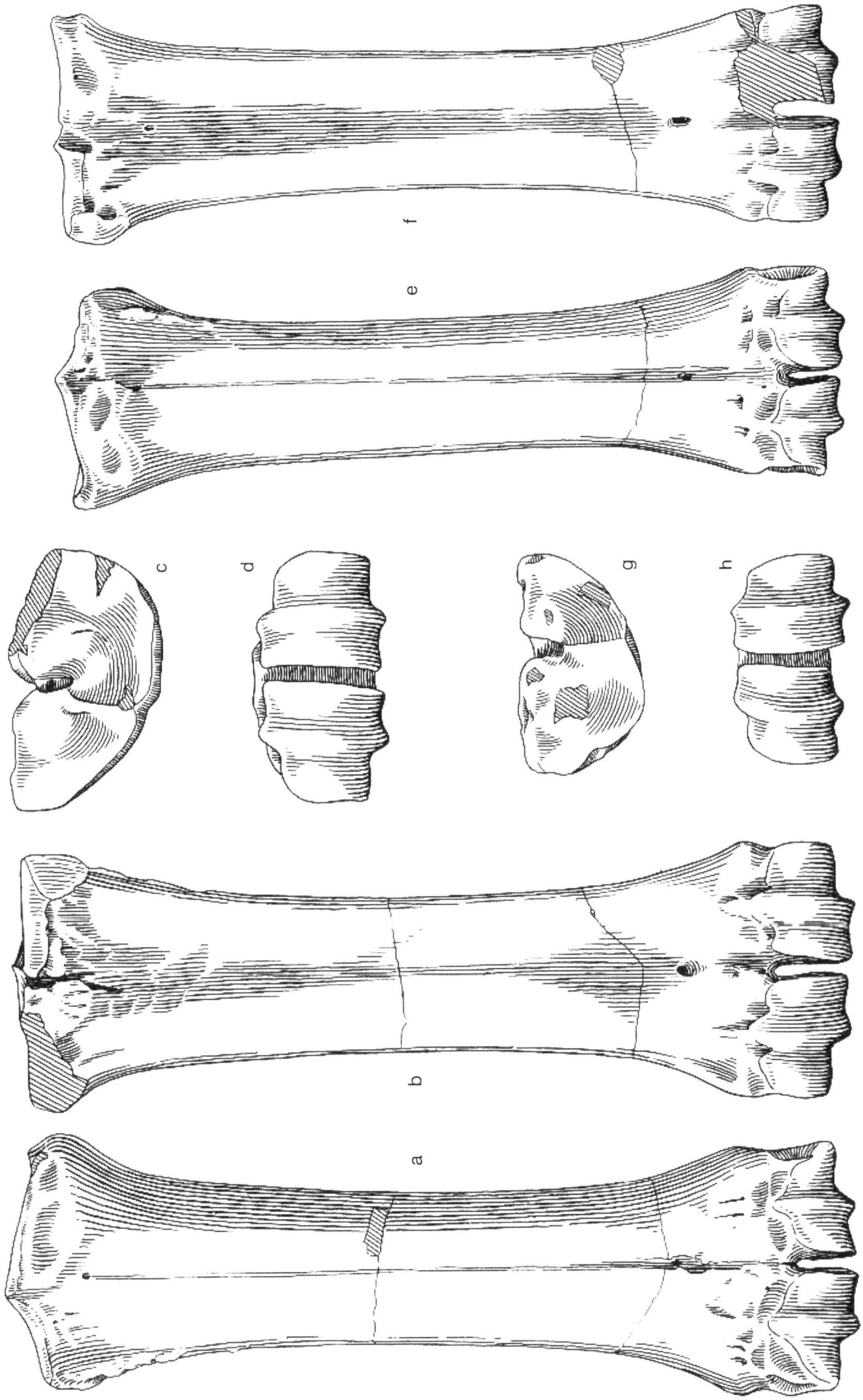

Fig. 5 *Bison menneri*, Untermassfeld.
Metacarpal bones: a-d Right metacarpal III+IV [IQW 1980/ 15 235 (Mei. 14 717)], male; anterior, posterior, proximal, and distal views. – e-h Left metacarpal III+IV [IQW 1980/ 16 658 (Mei. 16 179)], female; anterior, posterior, proximal, and distal views. Scale: approx. 1:2. .

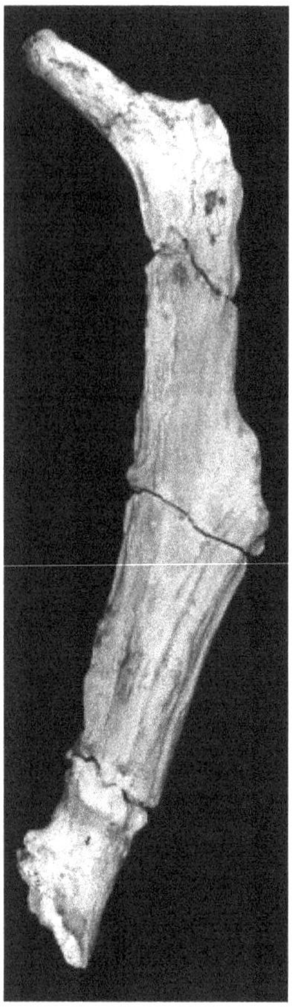

Fig. 6 *Capreolus cusanoides*, Untermassfeld. Left antler beam attached to skull fragment [IQW 1993/ 24 320 (Mei. 23 849)]. Scale: approx. 1:2.

Untermassfeld by finds of antler fragments as well as skulls and postcranial parts. Among the 47 individuals demonstrated within the entire body of material (MNI in Table 9), at least 19 adult males could be identified on the basis of antler remains (H.-D. Kahlke 2001a, 474). Meanwhile, four finds of more complete antlers have been made. They represent six-enders (H.-D. Kahlke 2001a, pl. 75-76, 1; Fig. 8 in the present publication). Leaving the attribution to genus indeterminate, H.-D. Kahlke (1997) referred this form to *Cervus* s.l. *nestii vallonnetensis*. It is to be assigned to a Villafranchian-Epivillafranchian group of forms of apparently Mediterranean origin, which following Azzaroli (1992) is often classified under the generic name of *Pseudodama*. Pfeiffer (1999) gave reasons for subsuming the Early to early Middle Pleistocene forms concerned under the genus *Dama*. She accordingly unites early fallow deer without palmate antlers at the subgeneric level as *Dama (Pseudodama)* and sets them against the late Middle Pleistocene to recent group of the fallow deer with palmate beams (subgenus *Dama*). A trend towards relative shortening of the premolar rows, as observed within the Late Villafranchian *nestii* "populations" of Italy, is continued in the stratigraphically younger finds from Untermassfeld. This feature, as well as a brow-tine which is set at a very low level, characterize the "small deer" from Untermassfeld as a late representative of the *nestii* group. Far-reaching morphological similarities exist between finds from Untermassfeld and the Grotte du Vallonnet (Alpes-Maritimes) (H.-D. Kahlke in Lumley et al. 1988, 483-484).

Finally, the faunal community of Untermassfeld includes very numerous remains of a large deer which may likewise be interpreted as a late representative of a group that had survived beyond the Villafranchian. Nearly 1,000 finds, a number similar to that recorded for *Bison menneri*, were available for study, but among them relatively few skull remains (e.g., Fig. 9) and antler fragments, some of which are recognizably attached to a skull fragment and show the characteristic features of *Eucladoceros* (Fig. 10). Finds of mature antlers that are less fragmentary are lacking. To date, only the antlered skull of a juvenile individual has been recovered (Fig. 11) (H.-D. Kahlke 2001a). This large and long-legged cervid was described as a new species, *Eucladoceros giulii* (H.-D. Kahlke 1997). Croitor & Kostopoulos (2004) referred it to the genus *Arvernoceros*. Solving this matter will probably only be possible when a complete antler is found.

The material of hippopotamids recovered at Untermassfeld is abundant and for the first time allowed the description and osteometric characterisation of almost every postcranial skeletal element (mostly represented by series) of a *Hippopotamus* "population" of the European Early Pleistocene (R.-D. Kahlke 1997b). In 1989, the then extant cranial and dental remains of juvenile individuals (on the basis of their milk dentition) were complemented by the skull of an adult animal, found broken into several parts (Figs. 12, 13). By the configuration of the facial skull bones and the demonstrably fully accomplished reduction of the incisors to the tetraprotodont mandibular dentition (R.-D. Kahlke 1985a), the finds class under the *amphibius* group of the African-European *Hippopotamus* branch (R.-D. Kahlke 1990). Subspecific rank is allowed in case of a separation from the recent form, and the name *Hippopotamus amphibius antiquus* is considered valid. Nonetheless, progressive characteristics of the skull construction mark a form which was evolved in comparison with the Late Villafranchian lectotype of *H. amphibius antiquus* from Figline (Upper Valdarno) (R.-D. Kahlke 2001e), but the morphology observed in stratigraphically younger skulls (*H. tiberinus* sensu Mazza 1991) were not reached at Untermassfeld.

The site of Untermassfeld has yielded only

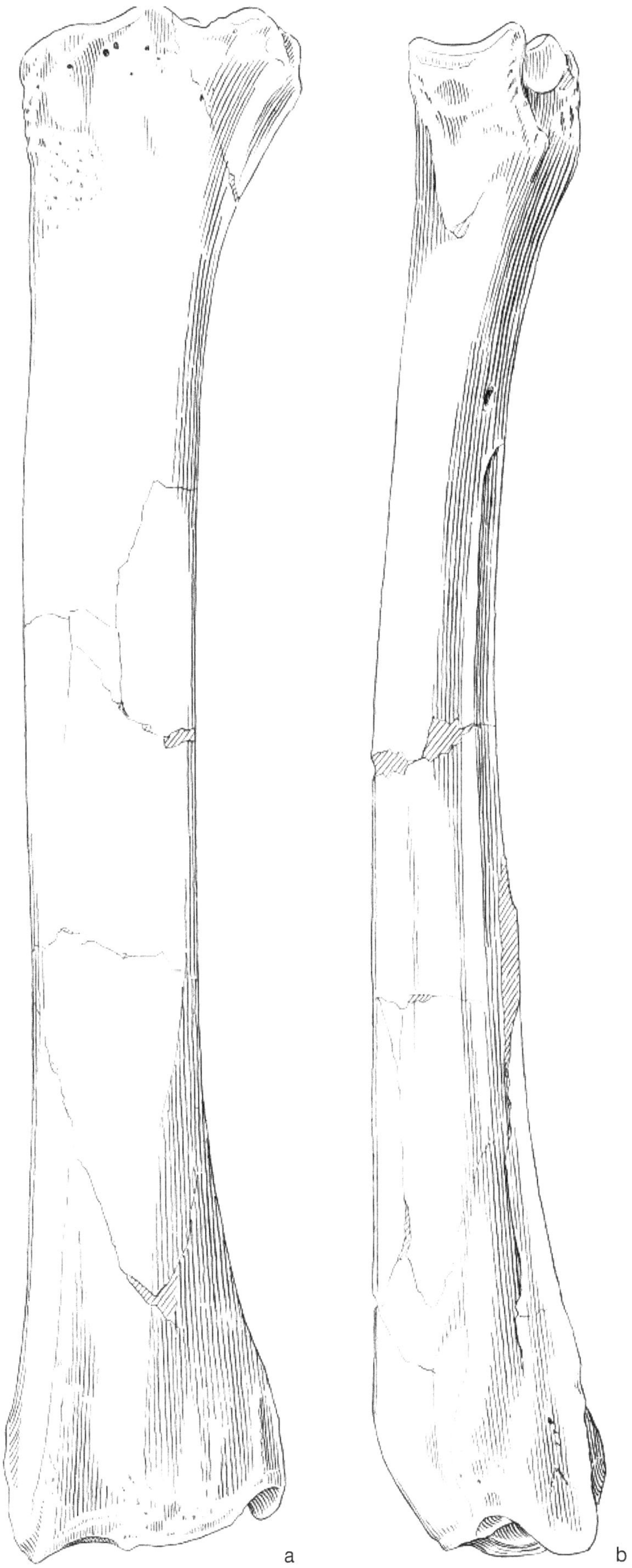

Fig. 7 *Alces carnutorum*, Untermassfeld.
a-b Left Radius/Ulna [IQW 1981/ 17 632 (Mei. 17 154)], anterior and lateral views.
Scale: approx. 1:2.

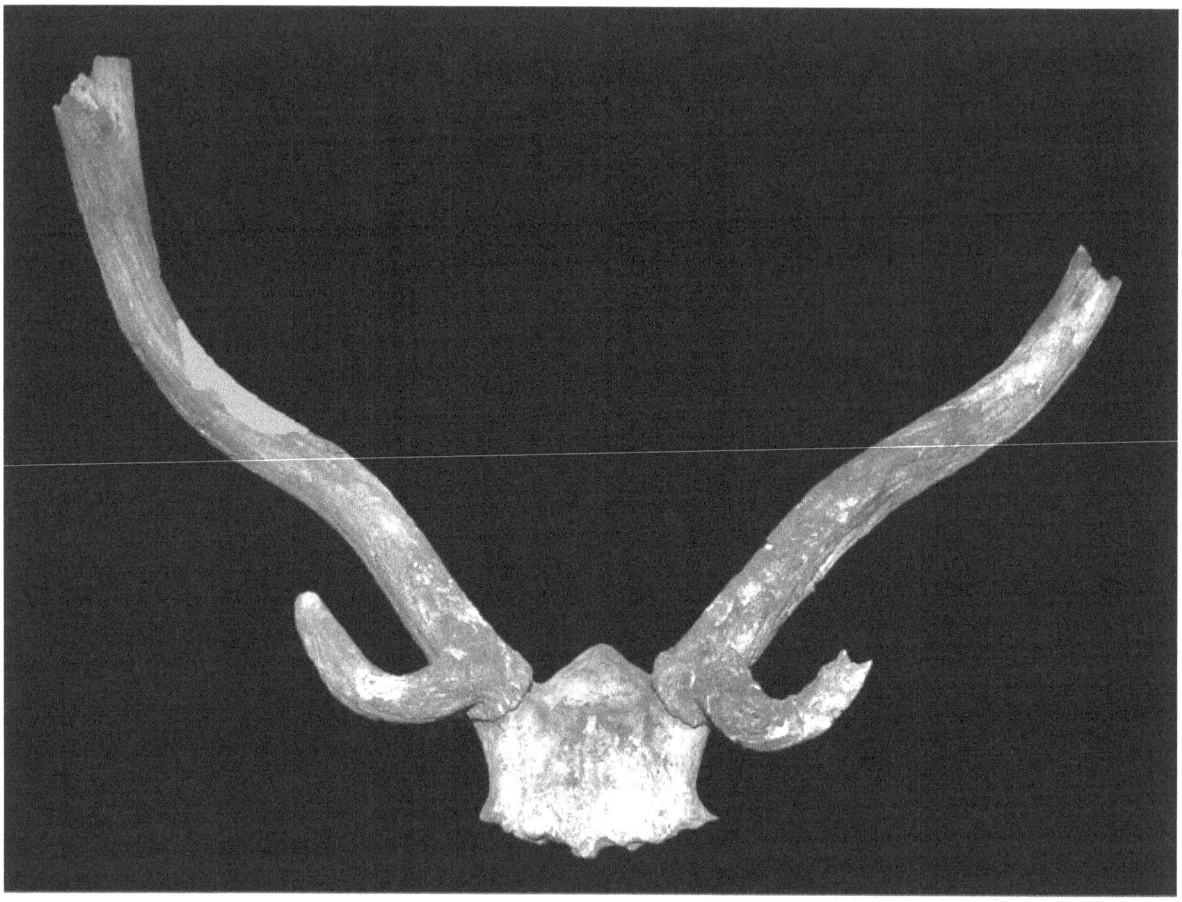

Fig. 8 *Cervus* s.l. *nestii vallonnetensis*, Untermassfeld.
Skull fragment with both antler beams [IQW 2003/ 28 185 (Mei. 27 347)]. Scale: approx. 1:4.

comparatively few remains of s u i d s, among them a juvenile skull with associated mandible. In addition to the specimens described by Guérin & Faure (1997), an atlas vertebra [IQW 1984/ 20 212 (Mei. 19 732)] was identified fairly recently. The material (Fig. 14) is attributable to *Sus scrofa priscus*, a large form that was characteristic of the late Early Pleistocene. Its molars are less complicated than those of the stratigraphically younger representatives of the genus.

Remains of r h i n o c e r o t i d s have been recovered in larger quantities (Fig. 15). Their study was based on approximately 650 specimens (H.-D. Kahlke 2001b). Within the evolutionary lineage of *Stephanorhinus etruscus*, the form from Untermassfeld is morphologically and metrically intermediate between the Late Villafranchian type material from the Upper Valdarno and

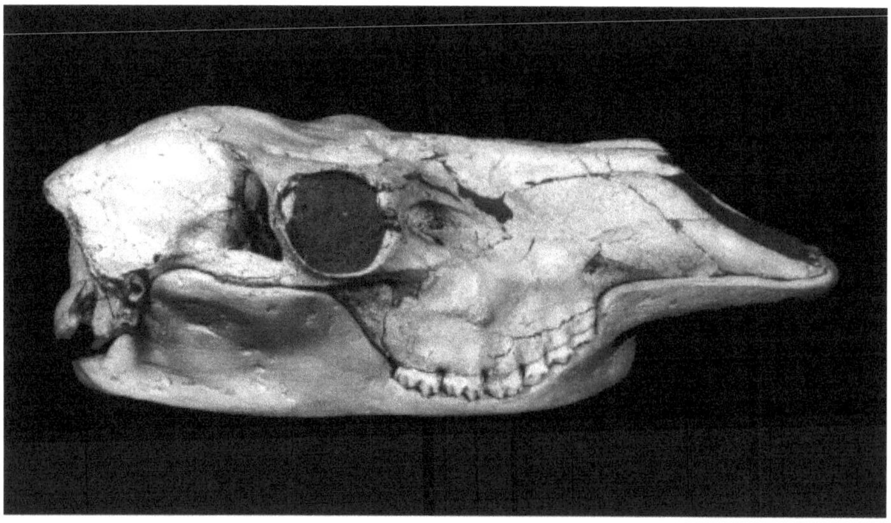

Fig. 9 *Eucladoceros giulii*, Untermassfeld. Skull with permanent dentition [IQW 1992/ 23 910 (Mei. 23 439)], female.
Scale: approx. 1:4.

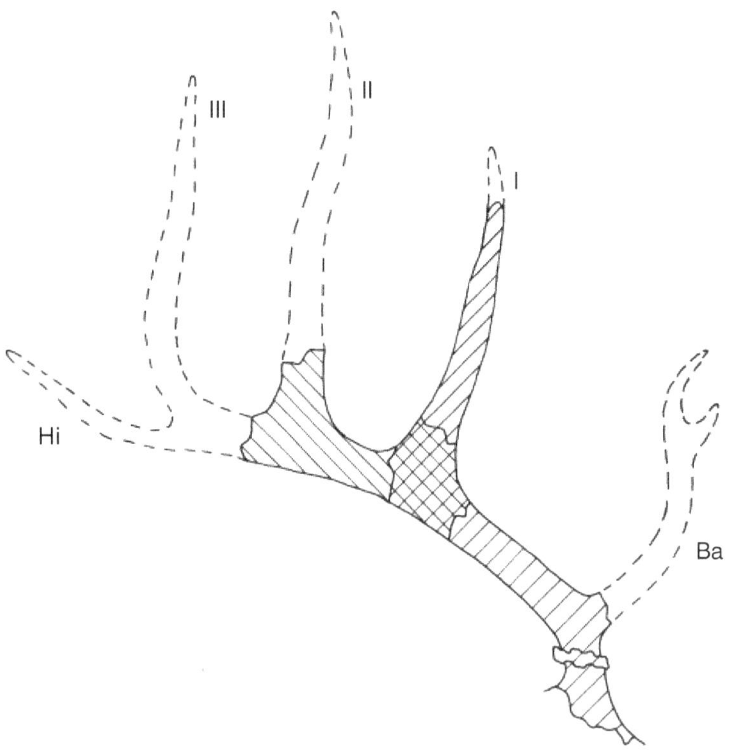

Fig. 10 *Eucladoceros giulii*, Untermassfeld.
Reconstruction of antler beam on the basis of the fragmentary specimens so far recovered: Ba Brow tine. – I-III Main tines. – Hi Posterior tine or end of beam (after H.-D. Kahlke 1997).

stratigraphically younger representatives (Fig. 16). A continuous size increase is documented from the Upper Villafranchian finds from southern Europe up to the later find complexes that belong to the Central-European 'pre-Elsterian' early Middle Pleistocene. This is made especially clear by the youngest northern populations (e.g., from Süssenborn). At first, separating Villafranchian and early Middle Pleistocene representatives of *Stephanorhinus etruscus* at the subspecific level on the basis of relative size differences is not difficult (*S. etruscus etruscus* and *S. etruscus hundsheimensis* = *S. etruscus brachycephalus*). However, establishing the evolutionary stage found at Untermassfeld is problematic, as it is intermediate both in shape and in body size. Hence it was not subsumed under one of the known subspecies, but a new subspecies name was not given either.

Remains of equids are only moderately abundant in the Untermassfeld fauna (Fig. 17). About 180 specimens could be included in this study, mainly postcranial skeletal elements and only a few remains of the dentition (Musil 2001a).

Fig. 11 *Eucladoceros giulii*, Untermassfeld.
Calvarium with antlers *in situ* [IQW 1990/ 23 672-75 (Mei. 23 201-04)]. Scale: approx. 1:3.

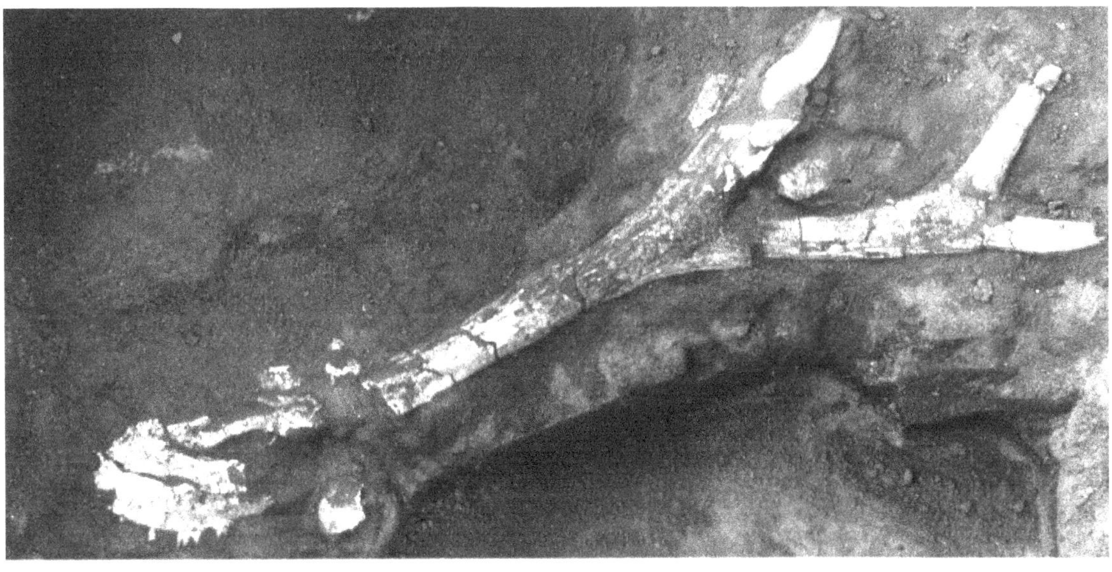

Some new finds that have hitherto not been taken into account were recovered especially during the 1996 field season. As far as known until then, all specimens were attributable to an average-sized species. In this respect Untermassfeld deviated from the majority of Villafranchian to early Middle Pleistocene sites in Eurasia, which normally document two sympatric stenonid equids that differ in body size (cf. Forstén 1988, 24-25). In July 1999, V. Eisenmann (Paris) was able to demonstrate the presence at Untermassfeld of a second, relatively robust equid on the basis of a single specimen (Eisenmann, in prep.). Morphologically more derived than comparable Villafranchian equids, the already described fossil "population" from Untermassfeld represents a form close to the stratigraphically younger *Equus altidens* (cf. also Forstén 1999, 1391). Musil (2001a) described it as a new species, *Equus wuesti*.

By contrast with other groups of mammals, only a small number of e l e p h a n t i d remains were available for study. Among these is only a single complete molariform tooth (D4 sin.; Dubrovo 2001). More recently a number of skeletal remains, including long bones, have been recovered, especially from the Channel Infill. In addition, a deformed skull of a juvenile individual (Fig. 18) with D^2-D^3 sin. et dex. [IQW 1996/ 25 646 (Mei. 25 175)] and fragments of lamellae of another pair of milk molars were found. Mainly on the basis of tooth morphology and the aserial build of the reconstructed carpus, Dubrovo attributed the extant material to the steppe mammoth, *Mammuthus trogontherii trogontherii*. It is likely, however, that it belongs to a late form of *Mammuthus meridionalis* (ssp.) (Lister et al. 2005, 54; cf. 7.2). The body height of a male adult from Untermassfeld was estimated at 4.4 m, that of a fully-grown female individual at 3.5 m (Dubrovo 2001).

The fossil community of Untermassfeld is distinguished by its extraordinary richness in carnivore material. Singularly complete remains of eleven species – with body masses mainly representing medium- and large-sized carnivore species – have so far been recovered. The absence of smaller carnivores cannot fail to attract attention. Mustelids, for example, are represented by only a single species (see below), and the presence of foxes could not be demonstrated at all. An interaction between ecological or ethological factors, and processes relating to the genesis of the site, could explain this phenomenon (cf. Sections 5.2., 6.1.).

Felids are represented in the faunal spectrum by six species of intermediate to large size. About 30 finds indicate at least five individuals of the European jaguar (*Panthera onca gombaszoegensis*) (Fig. 19). The material is morphometrically distinct from that of the late Middle to Late Villafranchian *P. onca toscana*, but falls within the variability range of the said younger European subspecies (Hemmer 2001; Hemmer et al. 2003).

A largely complete skull including the articulating mandible (Figs. 20b, 21), as well as four iden-

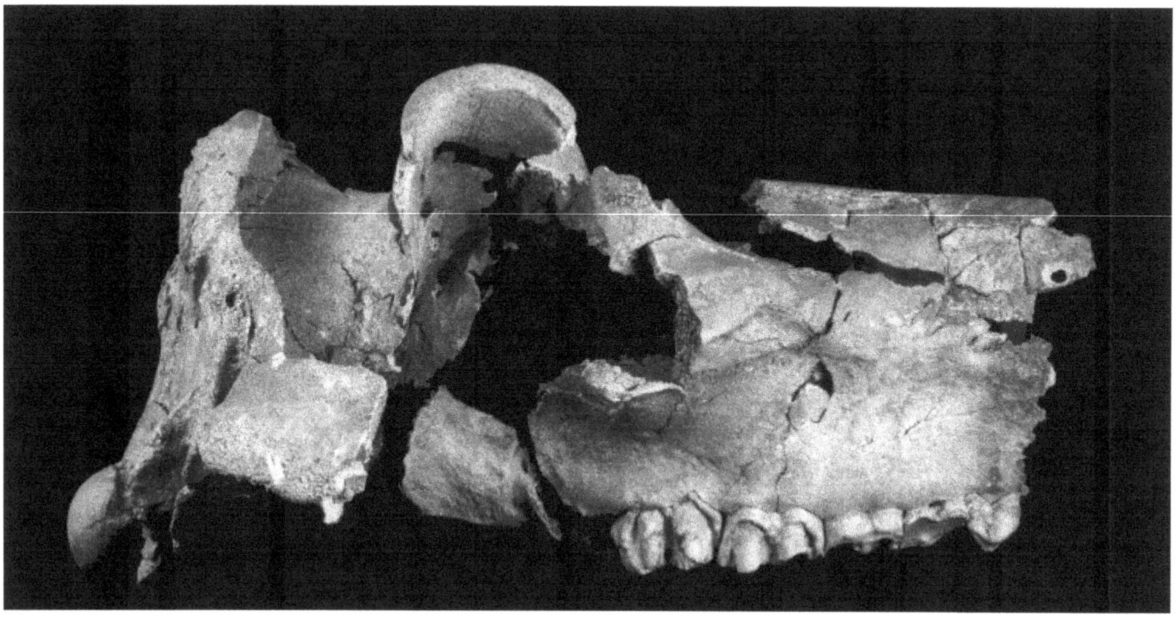

Fig. 12 *Hippopotamus amphibius antiquus*, Untermassfeld.
Mounted skull [IQW 1991/ 23 909 (Mei. 23 438)], slightly deformed. Scale: approx. 1:4.

tified cervical vertebrae, the right hind-limb, and two metacarpals of a single individual, belonged to an uncommonly large cheetah transitional between *Acinonyx pardinensis* and *A. jubatus*. Its name, *Acinonyx pardinensis pleistocaenicus*, takes into account the morphological proximity to Late Pliocene / Early Pleistocene finds from northern China (Hemmer l.c.). During the digging campaign of 2003 the find of a partial skeleton with complete skull produced evidence of a second, apparently somewhat smaller individual of *Acinonyx* (R.-D. Kahlke 2003b). Whereas the strongly deformed skull (Fig. 20a) is already prepared, the postcranial parts are still (2006) contained within

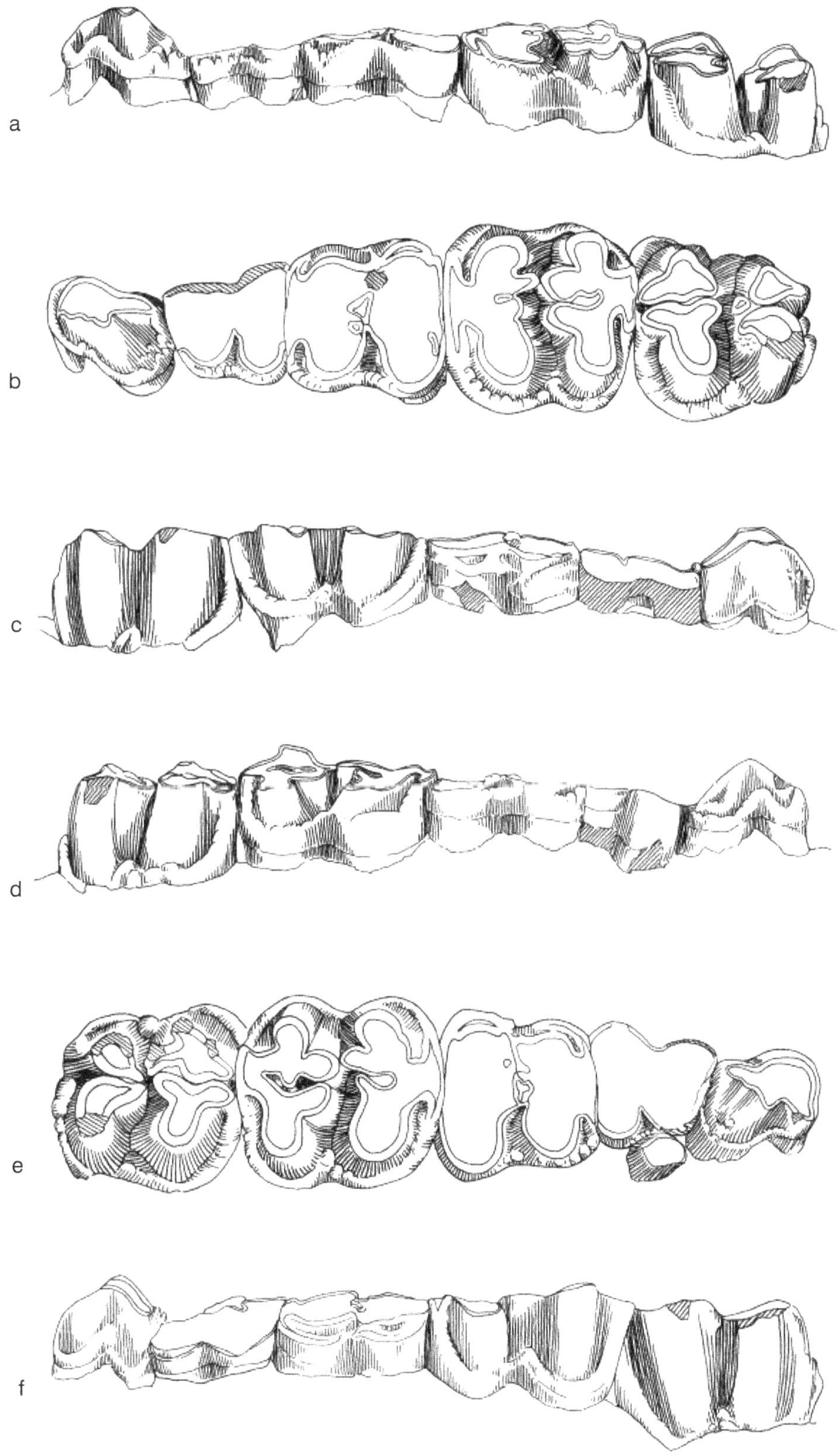

Fig. 13 *Hippopotamus amphibius antiquus*, Untermassfeld.
a-c Left maxillary tooth row with P^3, D^4, M^1-M^3 [IQW 1991/ 23 909 (Mei. 23 438)], lingual, occlusal, and buccal views. – d-f Right maxillary tooth row with P^3, D^4/P^4, M^1-M^3 [IQW 1991/ 23 909 (Mei. 23 438)], lingual, occlusal, and buccal views.
Scale: approx. 1:2.

Fig. 14 *Sus scrofa priscus*, Untermassfeld.
Left upper canine [IQW 1984/ 20 269 (Mei. 19 789)], male; superior, medial, and lateral views.
Scale: approx. 1:2.

a large, separately recovered block (R.-D. Kahlke 2003b; 2004a, 369).

A mandible fragment (Fig. 22) and some postcranial elements, including a pelvic fragment and long bones, are referable to at least one individual of a leopard-sized cat. Hemmer (l.c.) identified these specimens – as well as other European finds that had hitherto mostly been referred to *Viretailurus schaubi* – as remains of pumas (*Puma pardoides*). Likewise, Sotnikova (1976, 134f.) referred Asian material to *Felis (Puma)*. Comparison with the recent American form shows that the puma from Untermassfeld represents a less evolved stage. Basing themselves on the finds from Untermassfeld, Hemmer et al. (2004) drew up a survey of the now twelve Early Villafranchian to early Middle Pleistocene puma occurrences in the Old World.

Until now, remains of the lynx have only occasionally been recovered at Untermassfeld. A mandible with a dentition that lacks only the incisors (Fig. 23), a fragment of another mandible, and a humerus indicate that the animals are phylogenetically situated between European Late Villafranchian and Western European Middle Pleistocene representatives of the genus (Hemmer 2001). Taxonomically the lynxes from Untermassfeld are classified as *Lynx issiodorensis* ssp. ex aff. *spelaeus*.

As far as known, Untermassfeld represents the stratigraphically youngest occurrence of the dirktooth cat *Megantereon* in Europe. This form (Fig. 24) is equipped with an extremely reduced P^3 and was classified as *Megantereon cultridens adroveri* by Hemmer (l.c.). It is documented by dental remains of two individuals and by a humerus fragment.

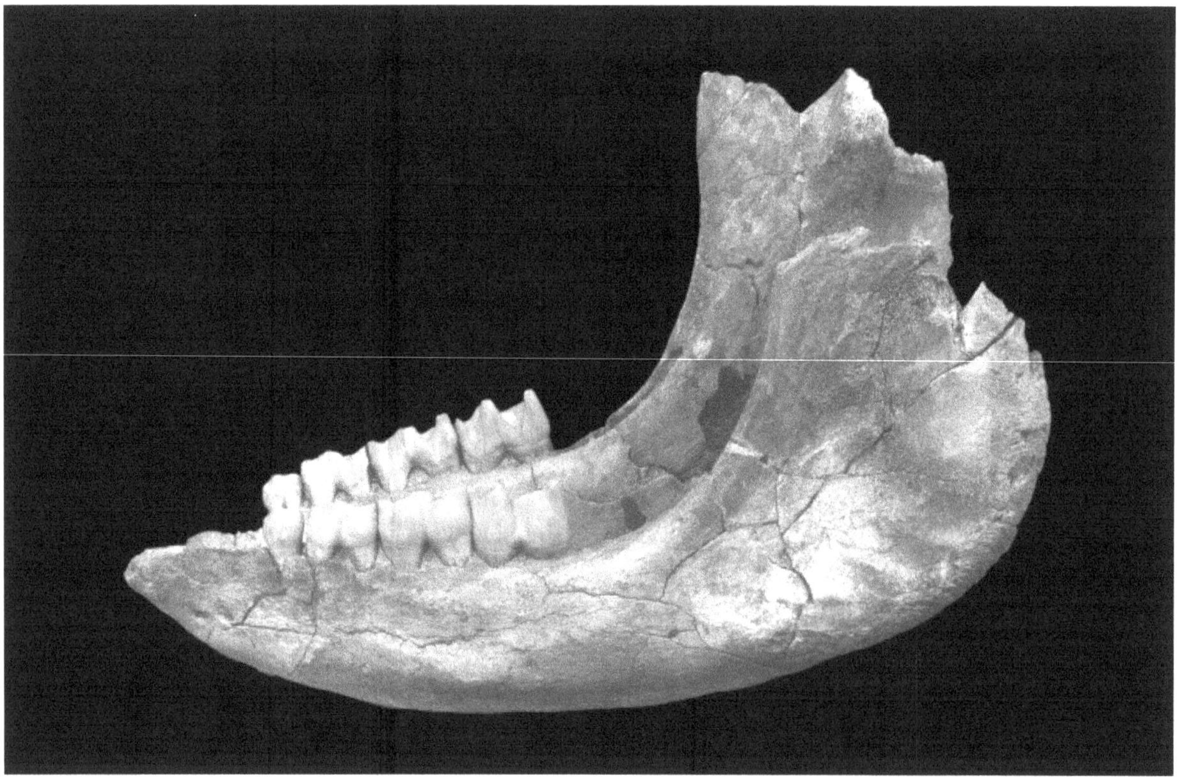

Fig. 15 *Stephanorhinus etruscus*, Untermassfeld.
Mandible with D_1-D_4, M_1 sin. and dex. [IQW 1992/ 23 932 (Mei. 23 461)]. Scale: approx. 1:3.

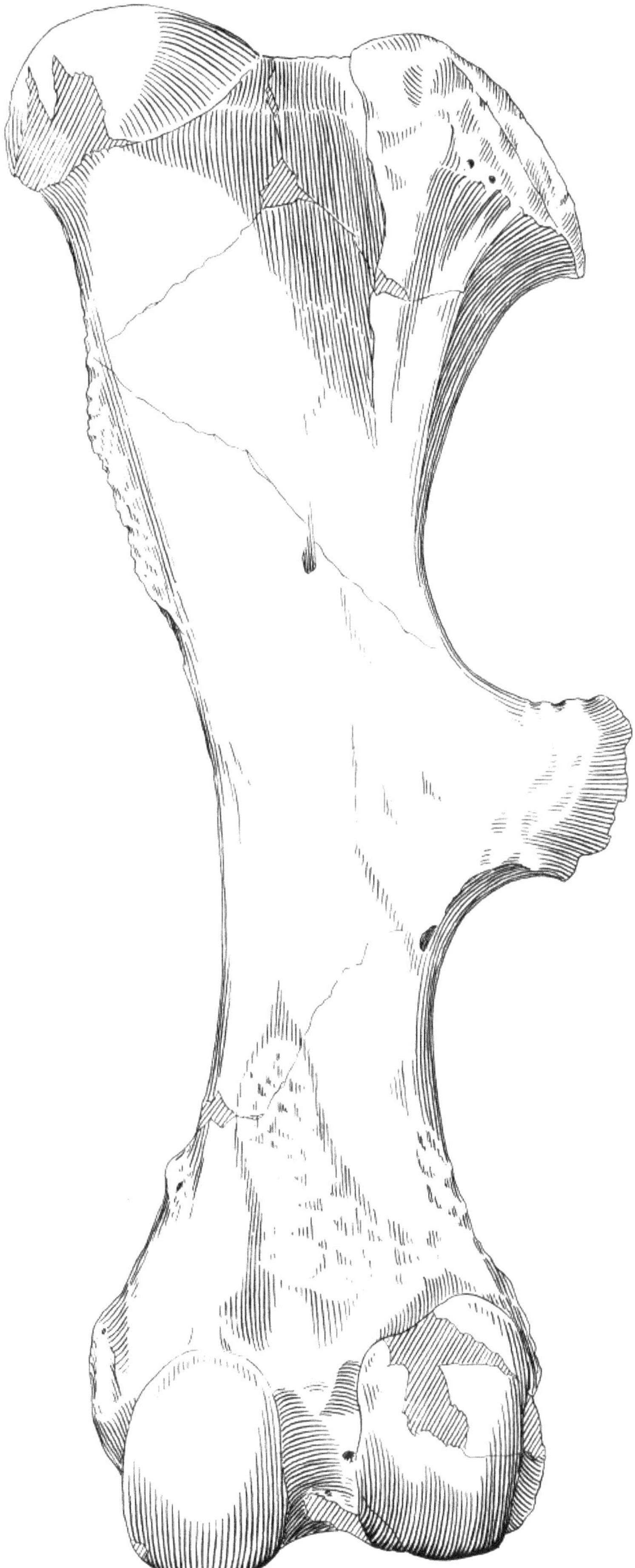

Fig. 16 *Stephanorhinus etruscus*, Untermassfeld.
Right femur [IQW 1980/ 16 047 (Mei. 15 558)], posterior view. Scale: approx. 1:2.

Fig. 17 *Equus wuesti*, Untermassfeld.
Partially preserved right hind extremity (centre) and additional skeletal elements *in situ* between specimens that belong to other species. Scale: approx. 1:6.

The largest felid in the Untermassfeld fauna is a sabretooth cat (*Homotherium crenatidens*) (Fig. 25). The coexistence of two morphotypes within a single population (the *crenatidens* and the *nestianum* type) provides positive proof of sexual dimorphism (Hemmer l.c.). A near-complete mandible (Fig. 26), isolated teeth, atlas and epistropheus, as well as single fragments of a humerus and a tibia indicate two large males and a smaller female.

To date, evidence of the European wild cat (*Felis lunensis*; cf. Hemmer 1993a, 1104-1105, 2003) has been lacking in the fossil material, although its presence is to be expected on the basis of both the time range of its occurrence and the ecological circumstances (cf. Sections 5.2., 6.). Fig. 27 gives an impression of the body shapes and size relations within the Untermassfeld felid community.

Nearly 100 skeletal remains from Untermassfeld are attributable to the very powerfully built hyaenid *Pachycrocuta brevirostris*. After the analysis of the available specimens had been completed (A. Turner 2001), additional finds have been made, among which is a skull of a juvenile animal. It was recovered in fragments and reconstructed [IQW 1996/ 25 647 (Mei. 25 176)] (Fig. 28). Dental remains are predominant within the material as a whole, but postcranial specimens are also present. Apart from the numerically preponderant juveniles, there is evidence of several adult individuals with largely complete sets of

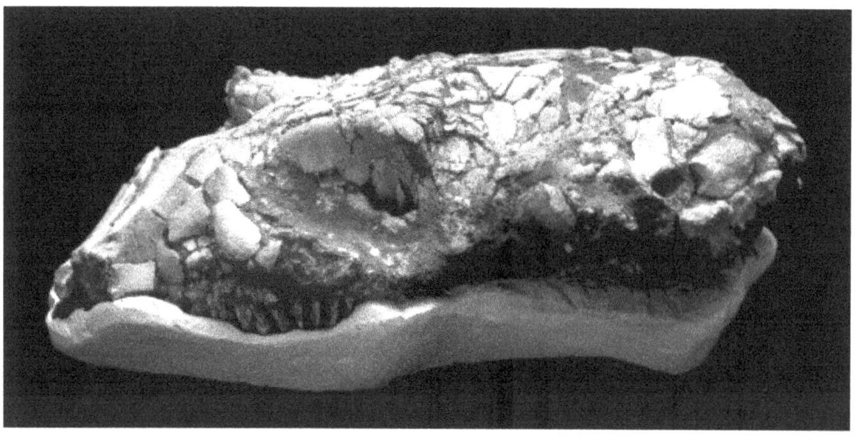

Fig. 18 *Mammuthus meridionalis* ssp., Untermassfeld.
Juvenile skull with D^2-D^3 sin. and dex. [IQW 1996/ 25 646 (Mei. 25 175)], *in-situ* mount. Scale: approx. 1:4.

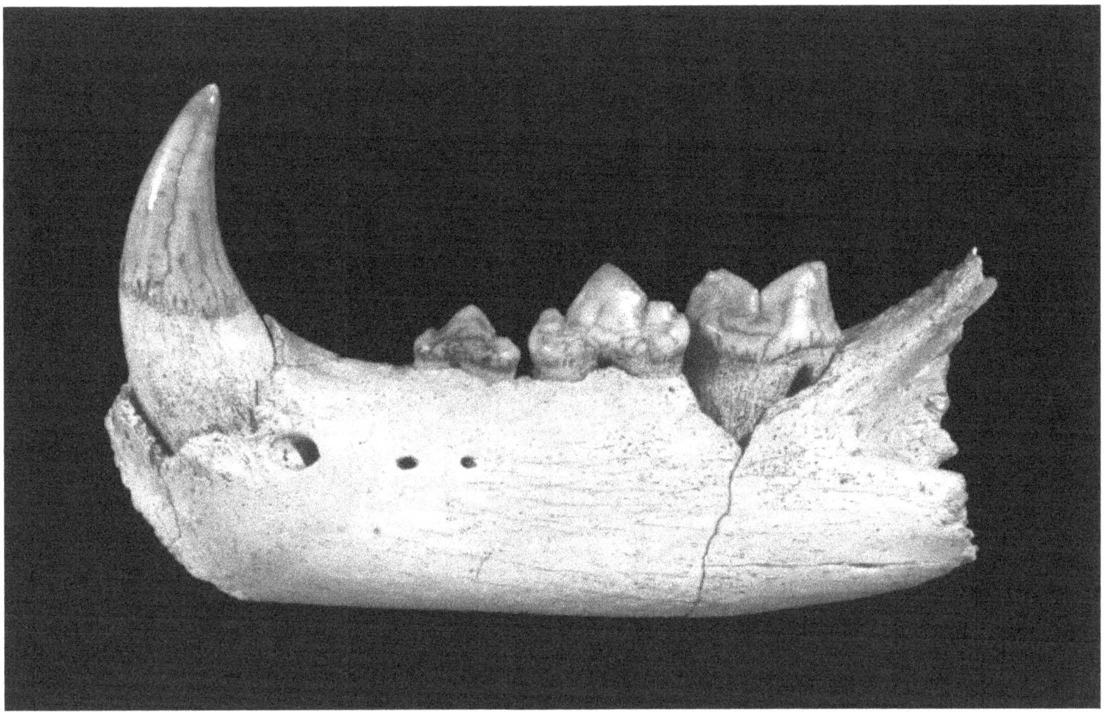

Fig. 19 *Panthera onca gombaszoegensis*, Untermassfeld.
Left mandibular fragment with C and P_3-M_1 [IQW 1983/ 19 169 (Mei. 18 689)], buccal view.
Scale: approx. 1:1.

teeth (Fig. 29). Metrical comparisons indicate the large dimensions of the body and especially the dentition of the species (cf. also A. Turner & Antón 1996). Frequently observed tooth marks on the bone material of nearly all large mammals represented at the site, as well as approximately 160 coprolites, bear witness to hyaena activities within the area of the present site.

Only three finds belong to the genus *Meles*, until now the only m u s t e l i d in the Untermassfeld fauna. They comprise fragments of a skull with some elements of the dentition (Fig. 30), as well as a left mandible ramus, referred by Wolsan (2001) to *Meles hollitzeri*, thus far known from late Early to early Middle Pleistocene horizons. Meanwhile, the material has been complemented by the M^1 sin. of the same individual [IQW 1994/ 24 548 (Mei. 24 077)].

With well over 100 specimens, the quantitative representation of u r s i d s is similar to that of hyaenas, although there are fewer individuals (Fig. 31). The representation of mandibular rami (Fig. 32) indicates a minimum number of nine individuals belonging to both sexes. The material available to Musil (2001b) had been recovered up to and including the campaign year 1988 (e.g., Fig. 33) and was referred to a new species, *Ursus rodei*. More recently, the analysis of new finds and a revision of the total body of material was undertaken by N. García. This shows that the bear from Untermassfeld shares a series of primitive characteristics with both the earlier *U. etruscus* and the recent *U. arctos*. It has, however, more points in common with *U. dolinensis*, described by Garcia & Arsuaga (2001b). Hence the population from Untermassfeld, together with the bears from Vallonnet and Atapuerca-Gran Dolina (TD4), may be attributable to a group of forms that was the starting point of the Western Palaearctic cave bear lineage (Garcia 2004).

Several bear remains that belong to the same individual, viz. a fragmentary skull with articulated mandible, eleven vertebrae (th 10-14, lu 1-6), and ten bones of the extremities, show an advanced stage of diffuse idiopathic skeletal hyperostosis (R.-D. Kahlke & Kierdorf 1997).

Typical of a series of relatively rich late Early to Middle Pleistocene mammalian sites of the Palaearctic is the sympatric occurrence of two c a n i d s that differ in body size (Sotnikova 1978, fig. 1, 1988). At Untermassfeld, the robust canid is *Canis (Xenocyon) lycaonoides*, the smaller form *Canis mosbachensis*. Sotnikova (2001) had two upper molars, three partly fragmentary mandibular rami and a humerus of *C. (Xenocyon) lycaonoides* at her disposal. In the meantime, the body of material has more than doubled through the identification of previous as well as new finds (Fig. 34). A total of seven mandibular rami now documents five individuals. Every possible long bone

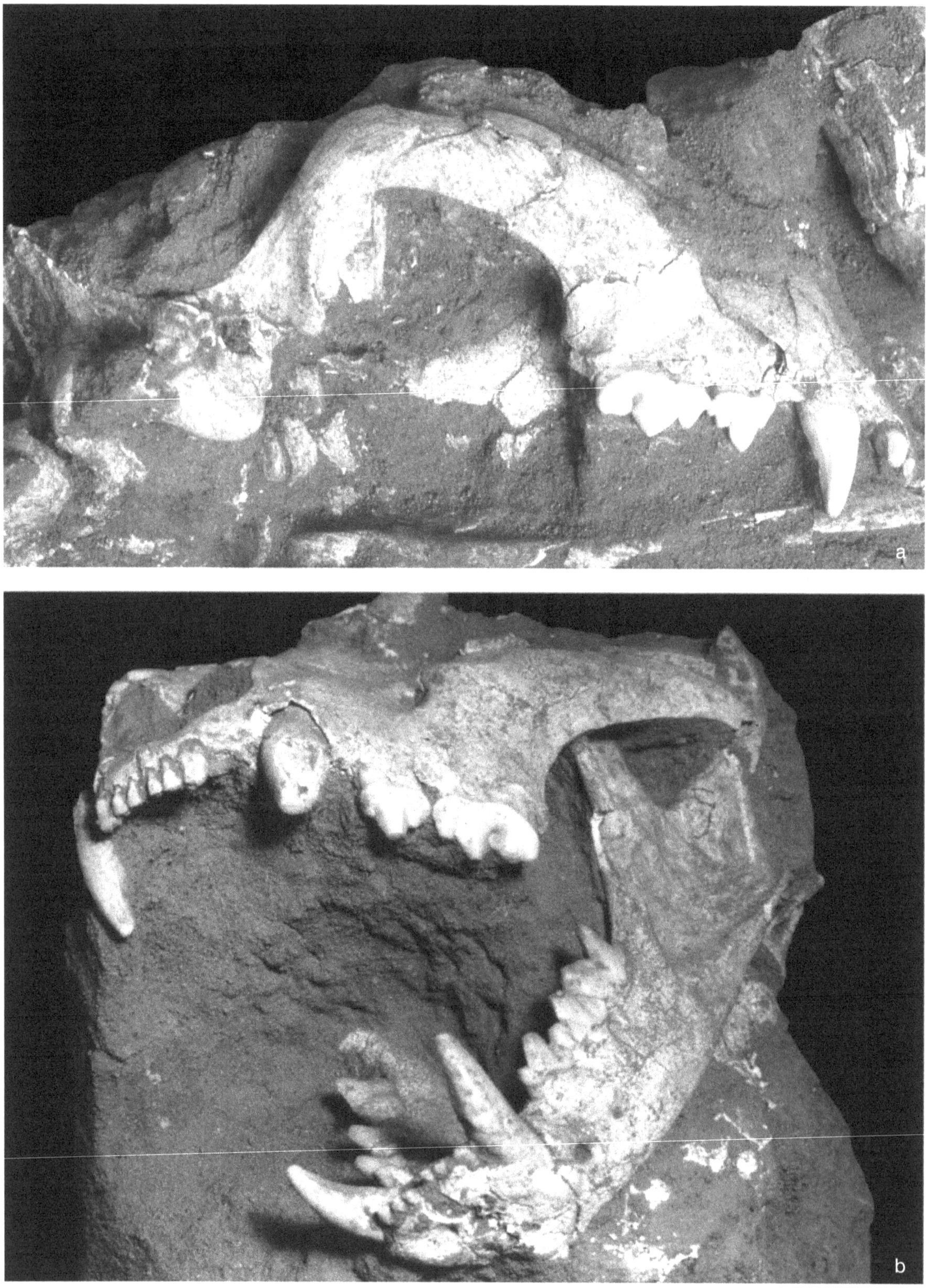

Fig. 20 *Acinonyx pardinensis pleistocaenicus*, Untermassfeld.
Complete skulls *in situ*. a Cranium [IQW 2004/ 28 616 (Mei. 27 778)], lateral view. – b Cranium with mandible [IQW 1980/ 16 350 (Mei. 15 861), IQW 1980/ 15 504 (Mei. 15 016), IQW 1980/ 15 503 (Mei. 15 015)], photograph upside down in relation to find-orientation. Scales: approx. 2:3.

of the body is present at least once and some of the specimens belong to the same individual. The material that belongs to *Canis mosbachensis* is extraordinarily rich and complete. More than 300 specimens were available for analysis (Sotnikova 2001), so that all essential skeletal elements could be examined. Among other new finds are a partial skeleton (digging campaign 1993) and another three skulls, which were found partly shattered (Fig. 35) (Keiler 2001b, pl. 68-69). According to its body dimensions, the "small wolf" from Untermassfeld is a medium sized canid. Sexual dimorphism is apparent in both the skull measurements and the postcranial skeleton (Fig. 36). Characteristics of the dentition and proportions of the extremities indicate an early, i.e., late Early Pleistocene, representative of the *mosbachensis* lineage.

To obtain the remains of small vertebrates, which are generally quite rare in fluviatile fossil

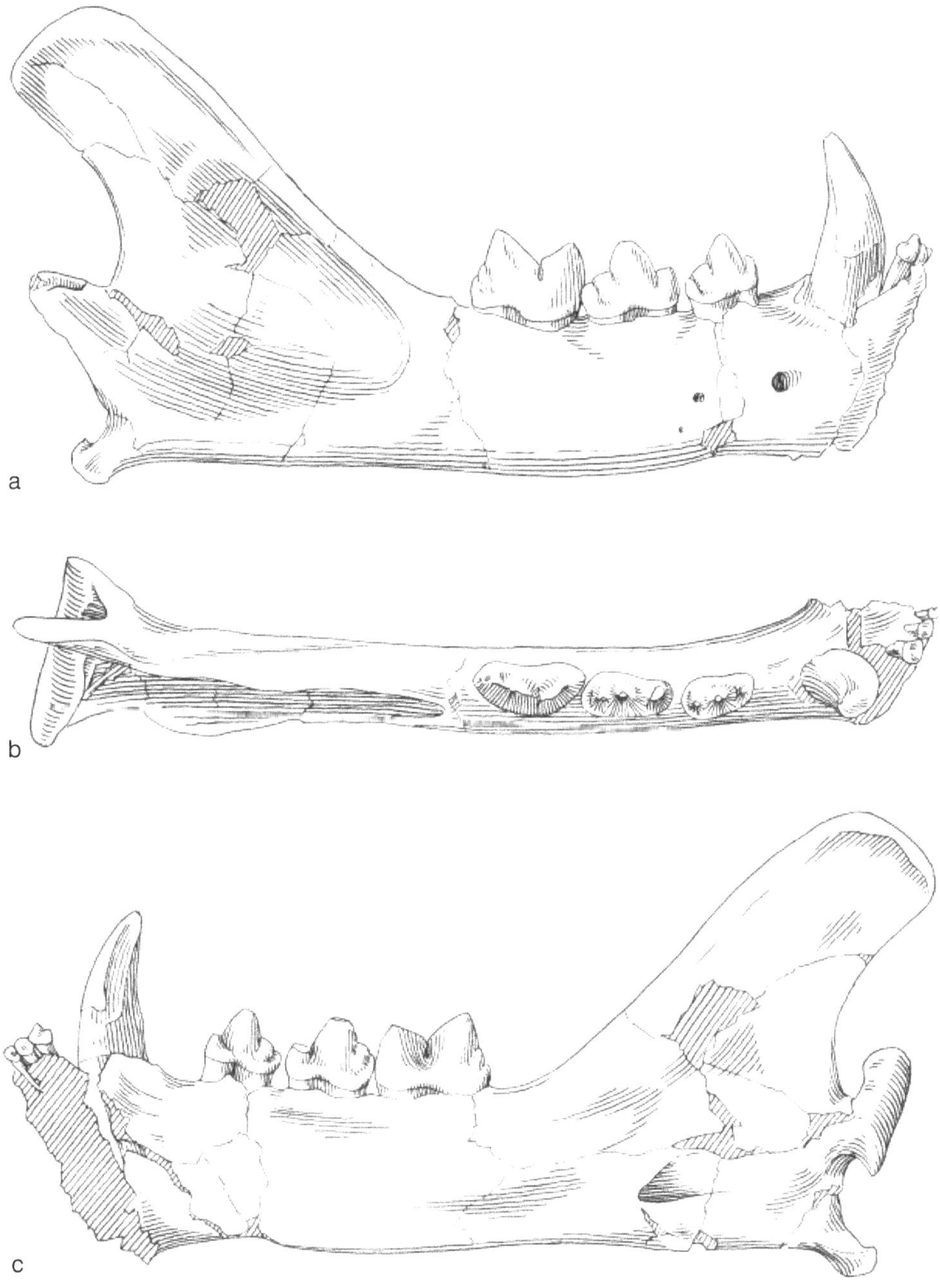

Fig. 21 *Acinonyx pardinensis pleistocaenicus*, Untermassfeld. a-c Right mandible half with complete dentition [IQW 1980/ 15 503 (Mei. 15 015)], buccal, occlusal, and lingual views. Scale: approx. 2:3.

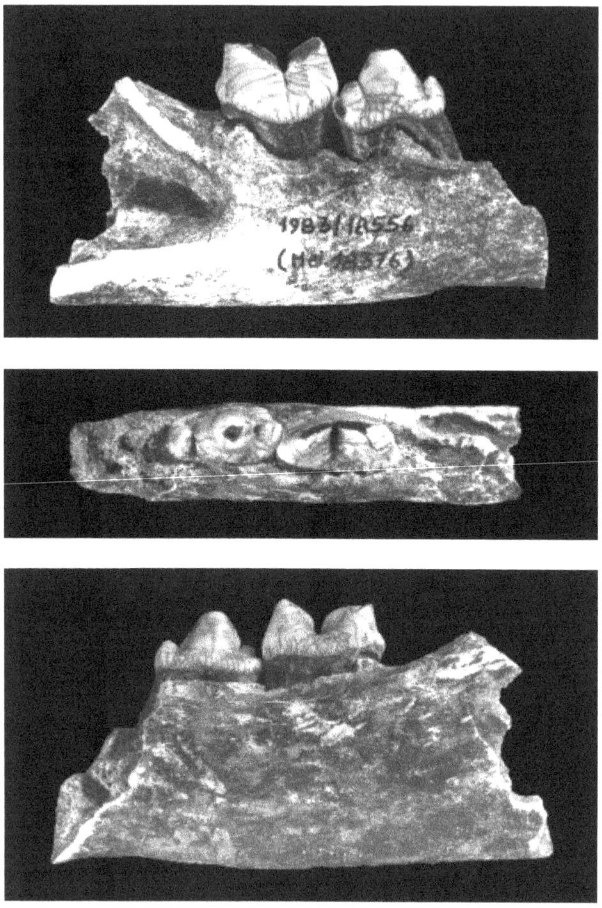

Fig. 22 *Puma pardoides*, Untermassfeld.
a-c Right mandibular fragment with P^4-M^1 [IQW 1983/ 18 556 (Mei. 18 376)], buccal, occlusal, and lingual views. Scale: approx. 1:1.

a

b

c

lagerstätten, large quantities of sediment - 170 tons in total – were washed and sorted over a period of 18 years. The r o d e n t s naturally dominate in the micromammalian material. Maul (2001a) identified 12 taxa in total. To begin with, some finds indicate the presence of a squirrel. Its molars are markedly smaller than those of the recent *Sciurus vulgaris* and motivate referral to *Sciurus* cf. *whitei*. The rodent most often represented in the recovered material is a ground squirrel. Because of the simultaneous expression of characteristics found in two species already described, the form is for the time being indeterminately referred to as *Spermophilus (Urocitellus)* ex gr. *primigenius / polonicus*. Beside abundantly represented isolated finds of single specimens (Maul 2001a), more or less incomplete partial skeletons of *Spermophilus* have regularly been recovered since the digging campaign of 2002, coming from an area directly under the lee of the Coarse Clastic Mudflow Fan that extends into the Lower Fluviatile Sands (cf. Sections 2., 4.3.1., 5.3.) (R.-D. Kahlke 2002b, 2003a). Currently there are eight incomplete skeletons and a near-complete one (Q 156, Fig. 37). The sediment structures in the vicinity of these anatomically connected finds do not indicate the existence of fossil burrows. The dead bodies of these small mammals were syngenetically washed in from the nearest surroundings.

The remarkable rarity of beaver remains among the material from Untermassfeld may be explained by the genesis model of the fossil lagerstätte (cf. Section 5.2.). Two mandibular rami and some isolated teeth document the presence of *Castor fiber*, and two additional tooth finds and an ulna dex. [IQW 2005/ 28 875 (Mei. 28 037)] belong to *Trogontherium cuvieri*. Apart from the gnawed facets which are effected by its lower incisors and exist on several bones of large mammals (Maul 2001b), two lower incisors produced evidence of the Old World porcupine (*Hystrix* sp.) (Figs. 38, 39).

The available teeth of a fat dormouse tally in size with those of *Glis sackdillingensis*. On the basis of tooth measurements, some hamster remains are attributable to *Cricetus* cf. *runtonensis* (Maul 2001a). In 2002 a disarticulated skeleton referable to this species was recovered (R.-D. Kahlke 2003a, fig. 3). Evidence of the bank vole

Fig. 23 *Lynx issiodorensis* ssp. ex aff. *spelaeus*, Untermassfeld.
a-b Mandible with C, P_3-M_1 sin. and dex. [IQW 1994/ 24 691 (Mei. 24 220)], occlusal and buccal views.
Scale: approx. 1:1.

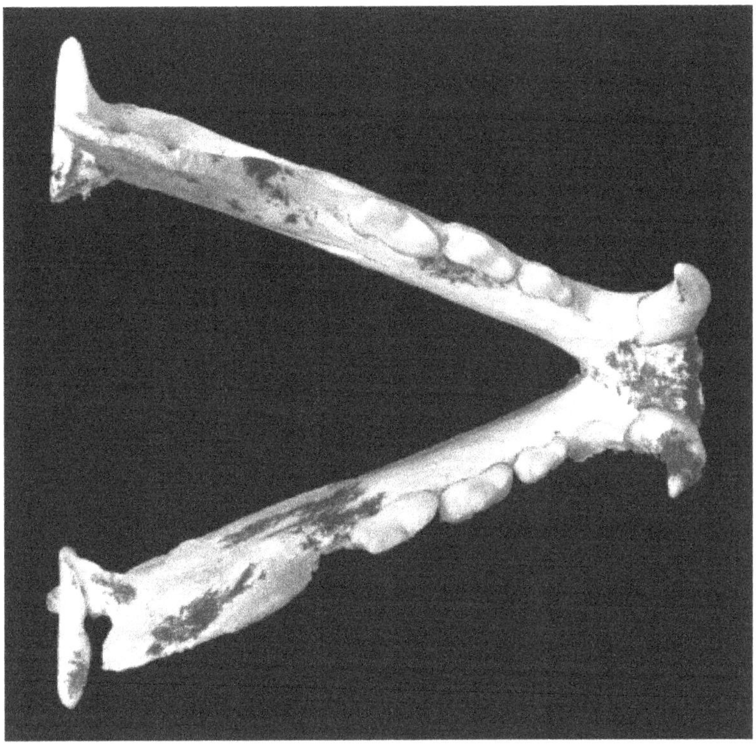

a

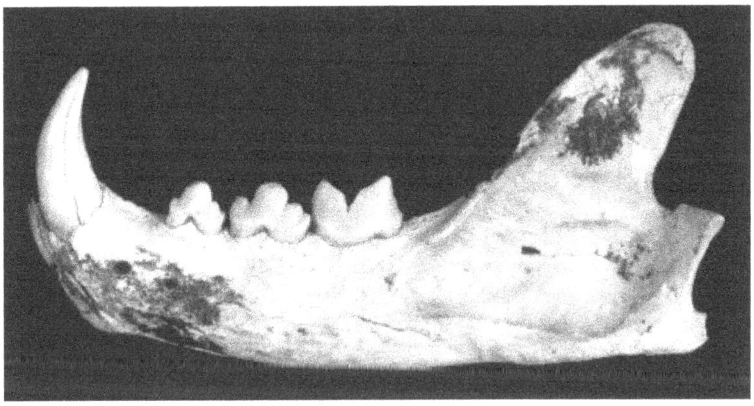

b

(*Clethrionomys* cf. *hintonianus*) is scarce. However, finds of the vole species *Pliomys episcopalis*, *Mimomys savini*, and *M. (Cseria) pusillus* have been somewhat more frequent, and *Microtus thenii* is represented quite often. In addition, some dental remains indicate the presence of a wood mouse (*Apodemus* cf. *sylvaticus*). Insignificant differences compared to the recent form arise from the robust build of accessory cusps on the M_1.

L e p o r i d remains count among the rarities in the Untermassfeld fossil assemblage (Maul 2001a). Until now there are only one upper premolar and two humerus fragments. According to their size, the latter may belong to either of the genera *Lepus* and *Hypolagus*. The premolar has to be assigned to a hare (cf. *Lepus* sp.).

The i n s e c t i v o r e s are represented by six species (Maul 2001a). A few finds per species document the hedgehog (*Erinaceus* cf. *europaeus*) and two mole species that differ in body size (*Talpa* cf. *europaea*, *Talpa* cf. *minor*). Furthermore, the shrews *Sorex runtonensis*, *Sorex (Drepanosorex)* ex gr. *margaritodon* – *savini* (transitional form) and *Beremendia* cf. *fissidens* are present.

Several separately recovered mandibular teeth of p r i m a t e s indicate at least one male and one female individual of *Macaca sylvanus* (Fig. 40) (Zapfe 2001). A subspecific attribution was not possible.

Within the entire body of extant material from Untermassfeld, hominid remains could not be identified, nor is intentional working of stone or bone objects demonstrable. The overall taphonomical results (cf. Section 4.) also do not suggest any anthropogenic influence.

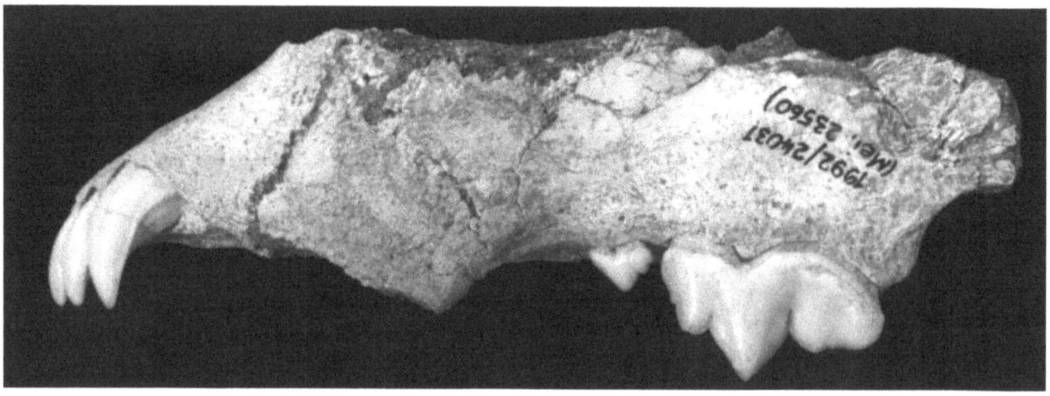

Fig. 24 *Megantereon cultridens adroveri*, Untermassfeld.
Left maxillary fragment with I^1-I^3 and P^3-P^4 [IQW 1992/ 24 037 (Mei. 23 560)], buccal view.
Scale: approx. 1:1.

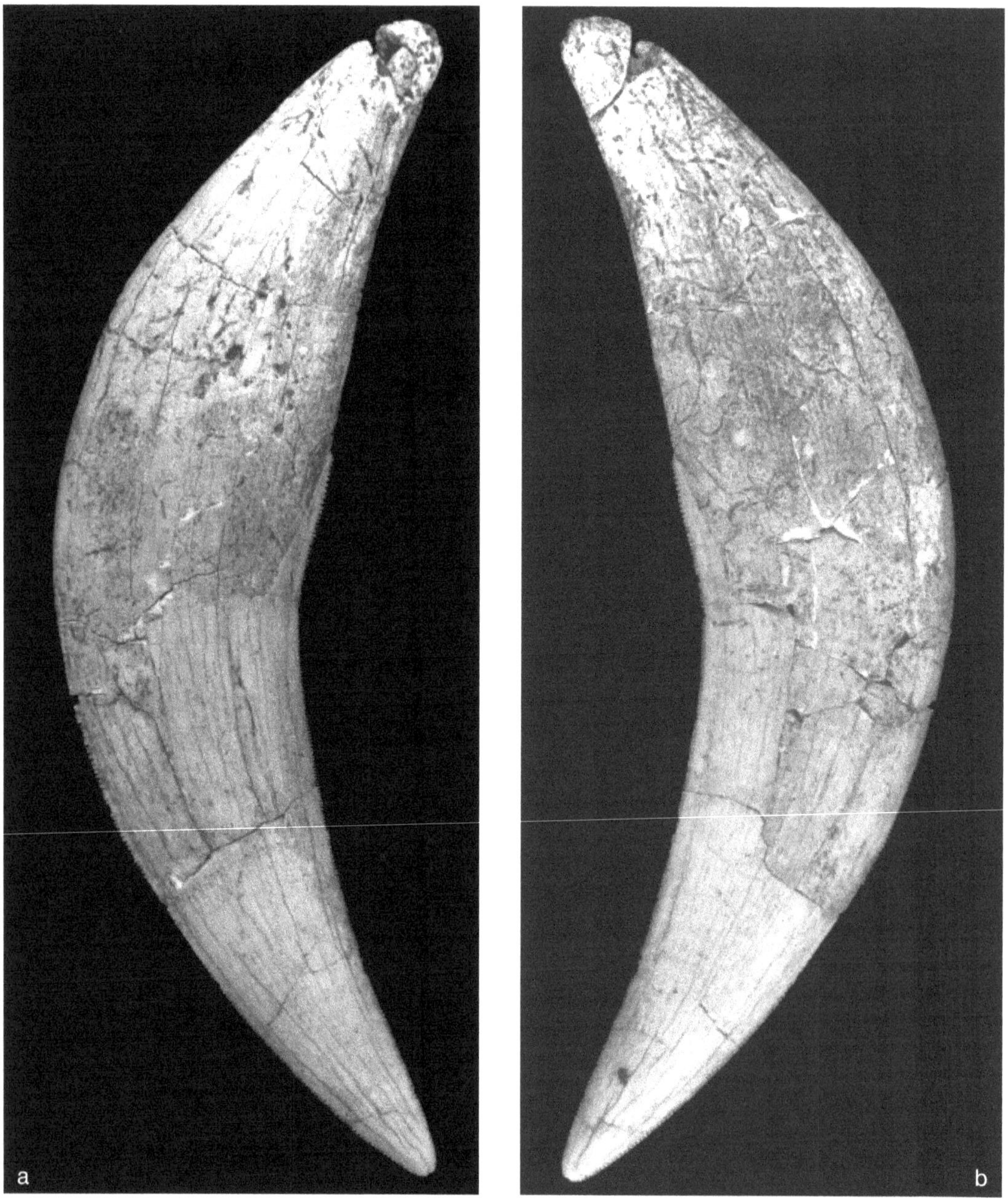

Fig. 25 *Homotherium crenatidens*, Untermassfeld.
a-b Right upper canine [IQW 1984/ 19 977 (Mei. 19 497)], lingual and buccal views.
Scale: approx. 1:1.

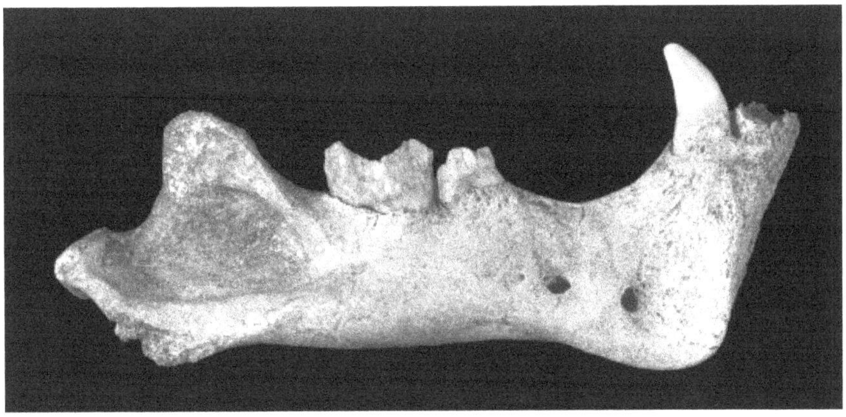

a

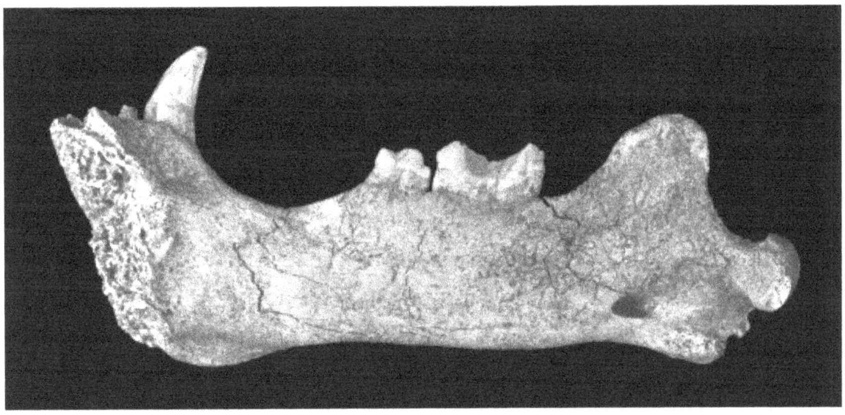

b

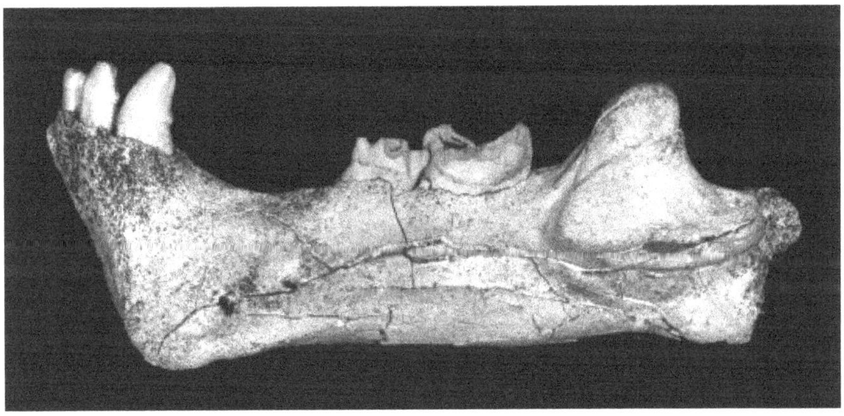

c

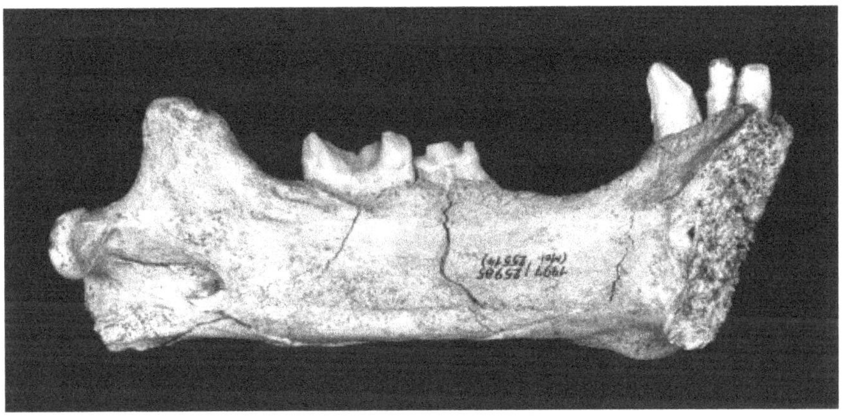

d

Fig. 26 *Homotherium crenatidens*, Untermassfeld.
a-b Right mandible half with C and P_4-M_1 [IQW 1993/ 24 372 (Mei. 23 901)], buccal and lingual views. – c-d Left mandible half with I_2-I_3, C and P_4-M_1 [IQW 1997/ 25 986 (Mei. 25 514)], buccal and lingual views.
Scale: approx. 1:2.

Fig. 27 The felid community of Untermassfeld.
Reconstructions to scale, relative sizes based on the finds from Untermassfeld: a *Acinonyx pardinensis*. – b *Homotherium crenatidens*. – c *Puma pardoides*. – d *Lynx issiodorensis*. – e *Panthera onca gombaszoegensis*. – f *Megantereon cultridens* (after Hemmer 2001).

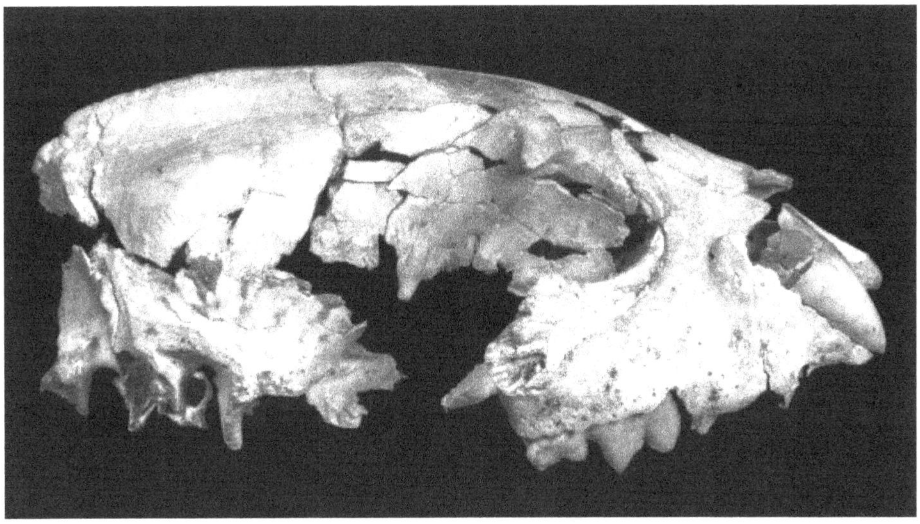

Fig. 28 *Pachycrocuta brevirostris*, Untermassfeld.
Juvenile skull in tooth replacement phase [IQW 1996/ 25 647 (Mei. 25 176)].
Scale: approx. 1:2.

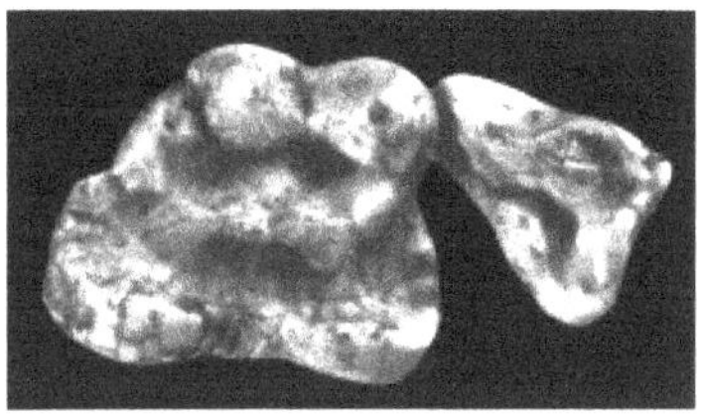

Fig. 30 *Meles hollitzeri*, Untermassfeld.
Right P^4 and M^1 from a partial skull [IQW 1994/ 24 619 (Mei. 24 148)], occlusal view.
Scale: approx. 2:1.

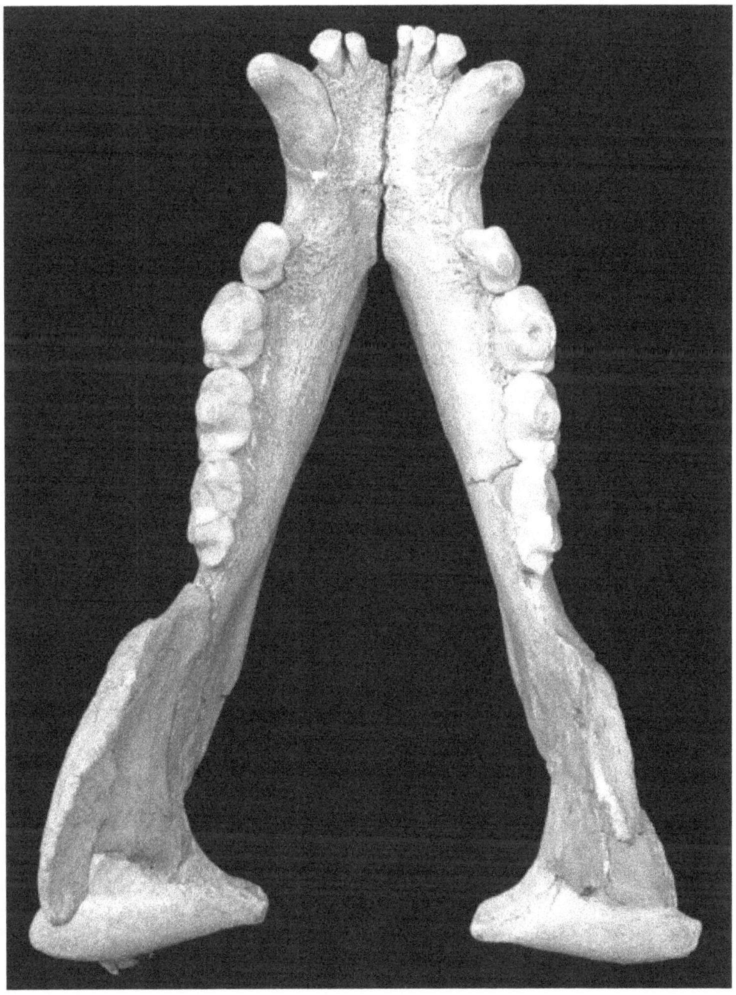

Fig. 29 *Pachycrocuta brevirostris*, Untermassfeld.
Mandible [IQW 1993/ 24 369 (Mei. 23 898), IQW 1994/ 24 584 (Mei. 24 113)], occlusal view.
Scale: approx. 1:2.

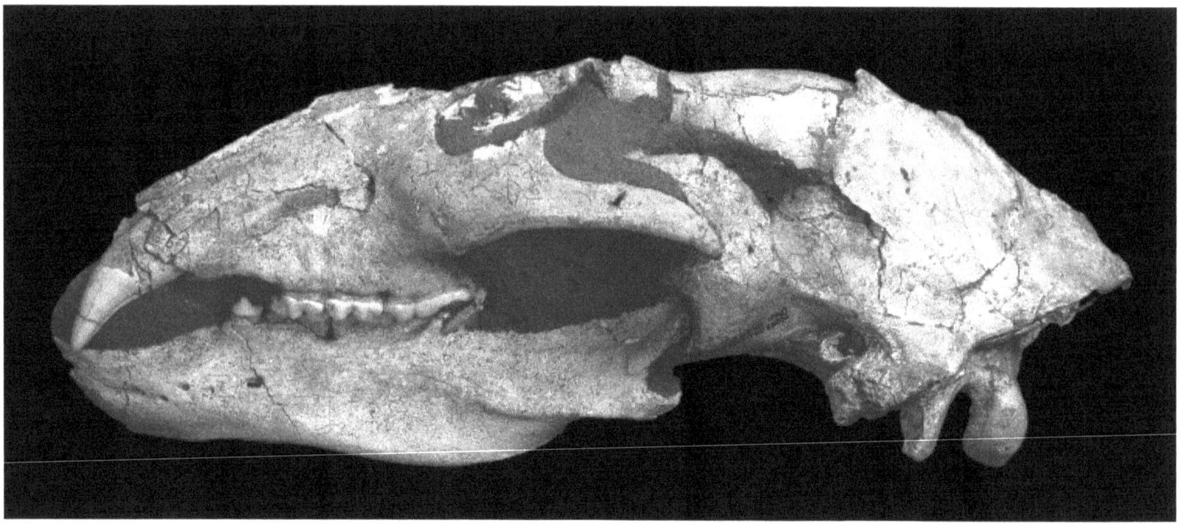

Fig. 31 *Ursus* cf. *dolinensis*, Untermassfeld.
Skull (with damaged brain case) and mandible [IQW 1995/ 24 844 (Mei. 24 373)]. Scale: approx. 1:4.

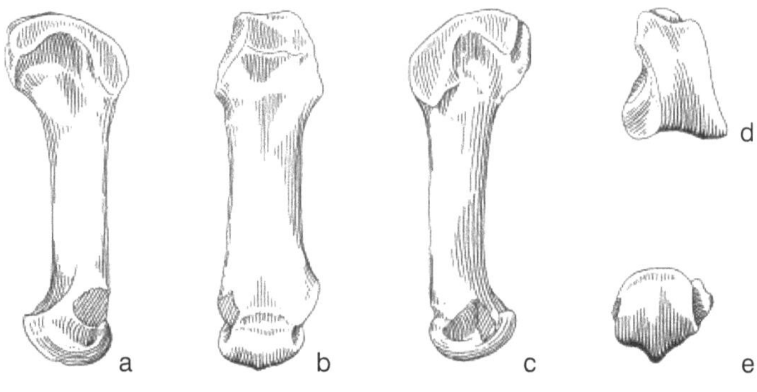

Fig. 33 *Ursus* cf. *dolinensis*, Untermassfeld.
a-e Left metacarpal III [IQW 1980/ 16 307 (Mei. 15 818)], medial, anterior, lateral, proximal, and distal views.
Scale: approx. 1:2.

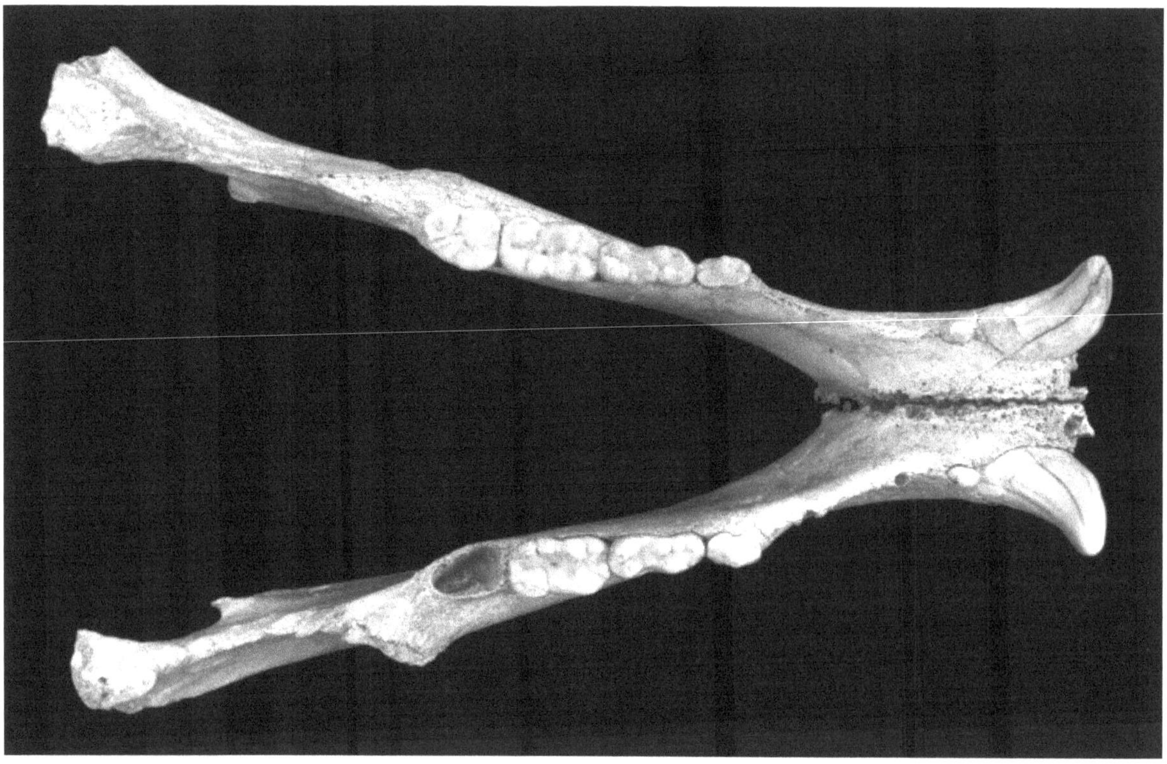

Fig. 32 *Ursus* cf. *dolinensis*, Untermassfeld.
Mandible [IQW 1980/ 15 522 (Mei. 15 034), IQW 1980/ 15 775 (Mei. 15 287)], occlusal view. Scale: approx. 1:2.

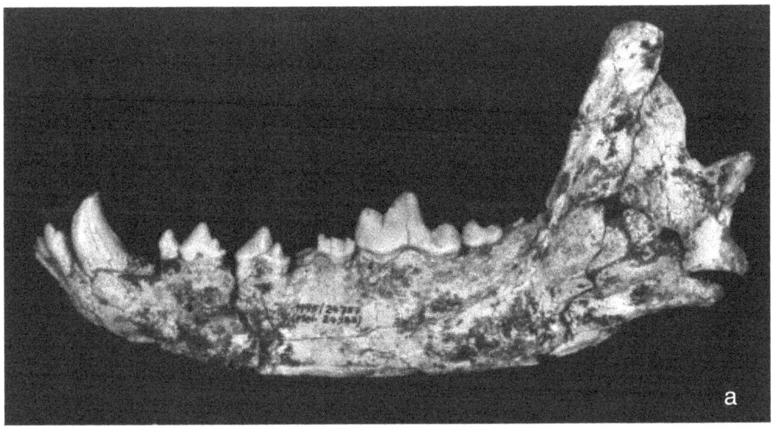

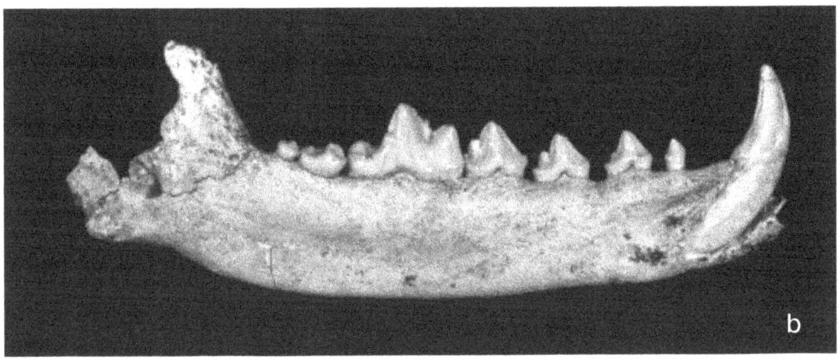

Fig. 34 *Canis (Xenocyon) lycaonoides*, Untermassfeld.
a Left mandible half with I_1-I_3, C and P_1-M_2 [IQW 1995/ 24 757 (Mei. 24 286)], buccal view. – b Left mandible half with C and P_1-M_3 [IQW 1993/ 24 371 (Mei. 23 900)], lingual view. Scale: approx. 1:2.

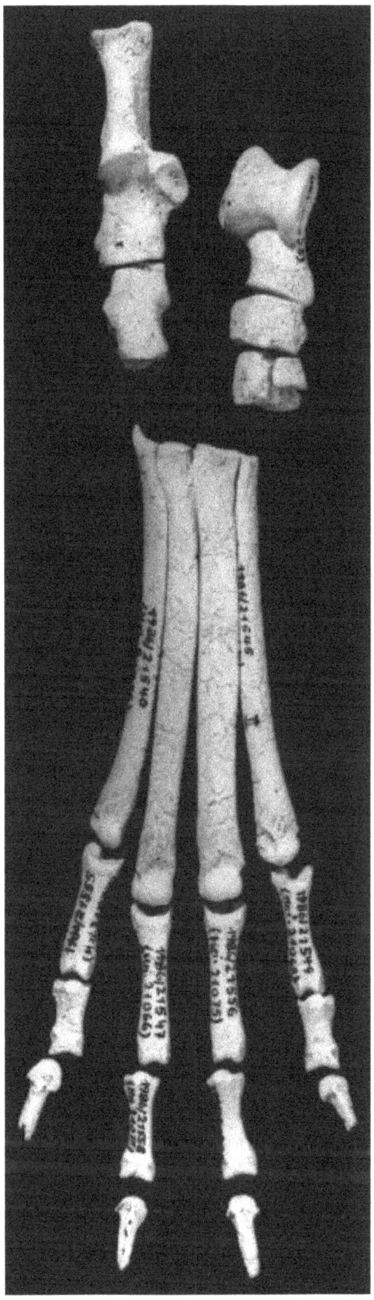

Fig. 36 *Canis mosbachensis*, Untermassfeld. Right hind foot, arranged from isolated finds. Scale: approx. 1:1.

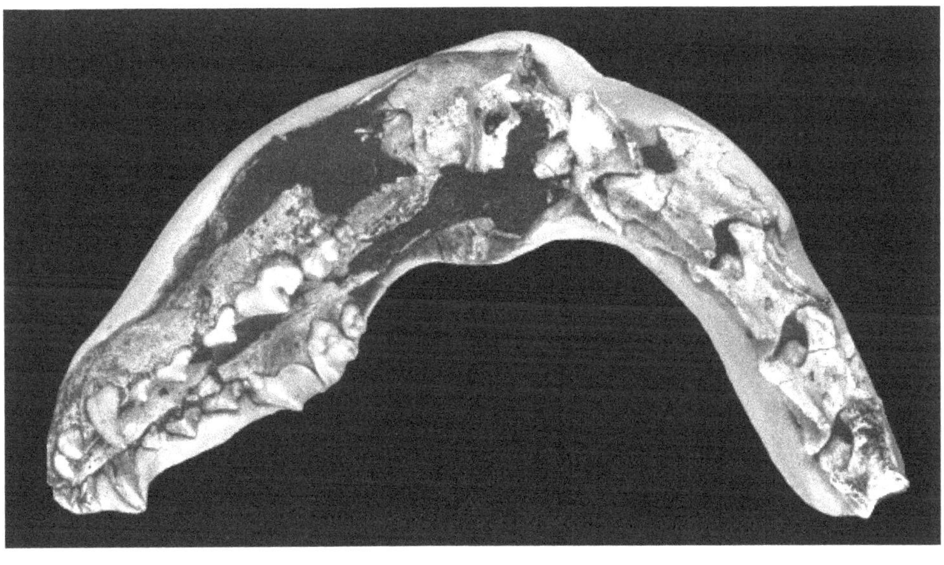

Fig. 35 *Canis mosbachensis*, Untermassfeld. Cranium and cervical vertebrae I-V [IQW 1997/ 26 017 (Mei. 25 546)], *in-situ* mount. Scale: approx. 1:2.

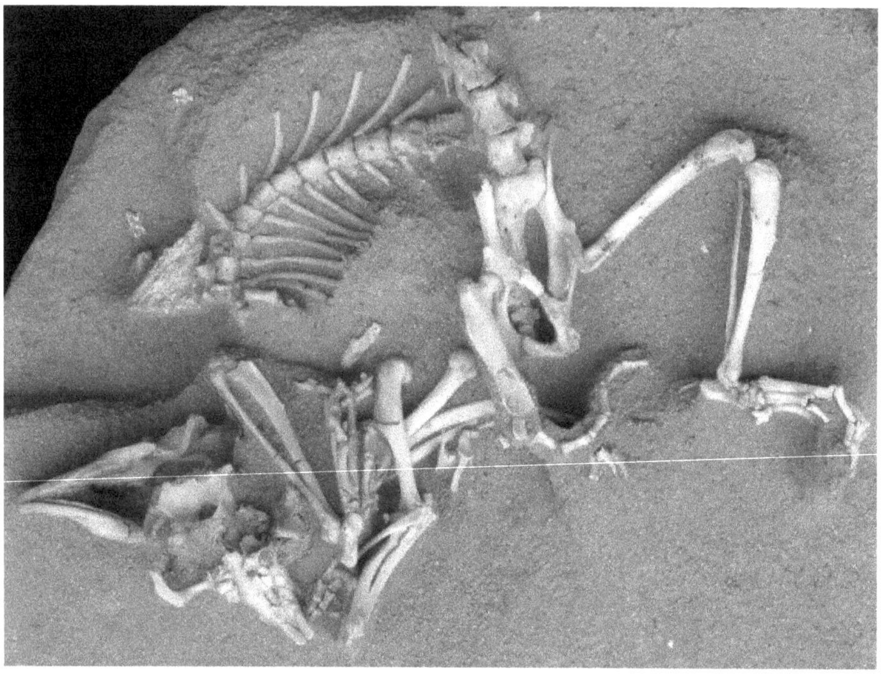

Fig. 37 *Spermophilus (Urocitellus)* ex gr. *primigenius / polonicus*, Untermassfeld. Articulated skeleton [IQW 2003/ 28 189 (Mei. 27 351)], *in-situ* mount. Scale: approx. 2:3.

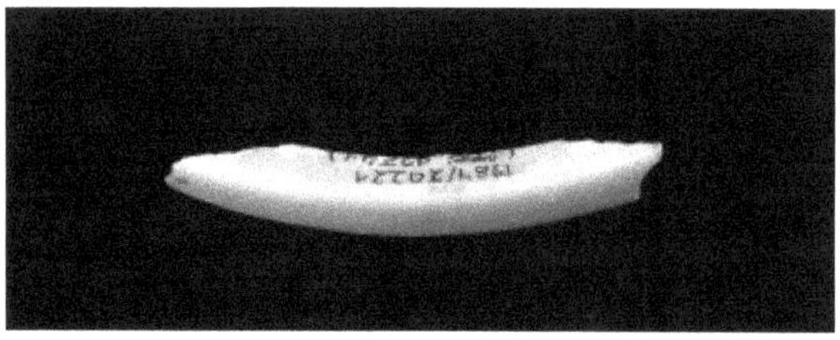

Fig. 38 *Hystrix* sp., Untermassfeld. Right lower incisor [IQW 1984/ 20 221 (Mei. 19 741)], lingual view. Scale: approx. 1:1.

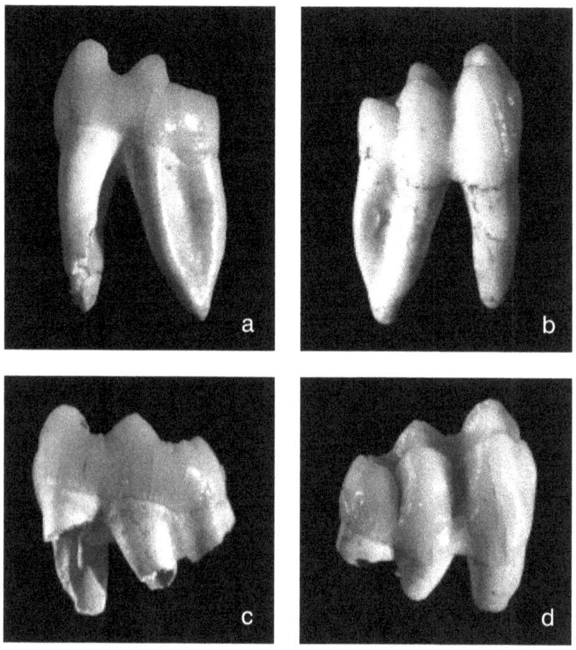

Fig. 40 *Macaca sylvanus*, Untermassfeld. a-b Right M_3 [IQW 1980/ 16 566 (Mei. 16 087)e], female; lingual and buccal views. – c-d Right M_3 [IQW 1984/ 20 021 (Mei. 16 541)], male; lingual and buccal views. Scale: approx. 2:1.

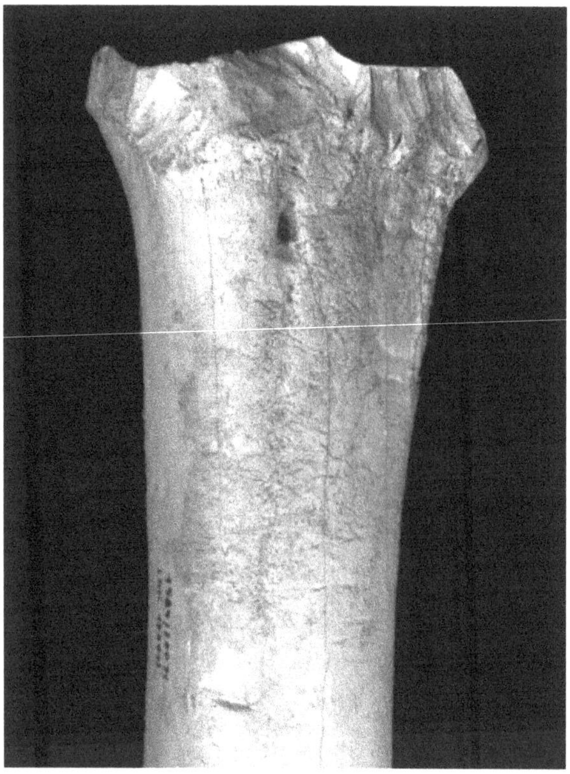

Fig. 39 Gnaw marks by *Hystrix* sp. on a metacarpal of *Bison menneri*, Untermassfeld [IQW 1984/ 20 071 (Mei. 19 591)], anterior view. Scale: approx. 1:2.

4. Taphonomy

4.1. Area of the excavation and documentation

The 584 squares (of 1 m^2 each) that were excavated within the Lower and Upper Fluviatile Sands between 1978 and the autumn of 1995 were mostly dug out to the level of the underlying floodplain sediment, or to the weathered coarse gravels in the area of the deepest part of the channel. Fig. 41 shows the configuration of the examined area and the rough distribution of the large mammal remains. No finds at all were indicated within the squares in the central part of the excavation area (roughly between Q 462, Q 118, Q 609, and Q 547). These represent the locations where loss of finds occurred before the presence of the assemblage was reported, and also undocumented areas from the time of the salvage excavations (cf. R.-D. Kahlke 1997b), which yielded a total number of approximately 1700 specimens in a predominantly excellent state of preservation.

In the course of the systematic excavations, all identifiable, analysable, remains of large mammals were three-dimensionally measured *in situ* and graphically recorded on field sheets[1]. Via the intermediate stage of so-called exact field sheets[2] – which also represent those finds in complete outline which were recovered *en bloc* – the present excavation plans were drawn up as definitive sheets and added to the present text as supplements (I-XV). Their respective positions within the excavated area can be gathered from Fig. 42. The plans show the specimens that were recovered at Untermassfeld up to and including 1995. All finds that had properly been measured received depth-related class marks. Relative to a fixed datum level (1.0 m above the main plane; cf. also the geological documentation in Ellenberg & R.-D. Kahlke 1997, and in R.-D. Kahlke 2001d), a height interval of 3.65 m [2.45 m below (Q 784) to 1.20 m above datum level (Q 922)] was excavated. Darker class marks indicate find positions at increasing depths.

The salvage excavations of the years 1979-1981 were recorded on field sheets that show the depth levels of the finds in relation to the terrain surface at that time. Recalculations of the find depths with respect to the datum level – which was not defined until 1983 – could afterwards not be exact. For this reason the contours of the bone remains concerned were rendered in broken lines (excavation plans I-VII). Exact data on the find depths are therefore missing here, but the positions within the excavation area are known. Specimens characterized in that way were mostly situated between 0 and 1.0 m below datum level. The plan squares to which this relates are also separately marked in Fig. 41[3].

The excavation plans added as enclosures show projections of all large mammal remains registered within the respective areas. Especially in the deepest part of the channel (excavation plans VII, X, XI), and within some other bone concentrations (excavation plans IV-VI, XI-XIV), a high density of finds led to a covering of the specimens underneath. In each specific instance, several detailed exact field sheets that graphically resolve the situation are additionally available for special case studies (see footnote 2).

4.2. Types of material preservation

The material preservation of the vertebrate fossils, after they had been covered by sediment, was predominantly determined by the geochemical and hydrogeological, and to a lesser degree also by neotectonic, circumstances at the site. The influx of carbonate-rich water from the Muschelkalk profile that outcrops upslope caused a slight impregnation of the entire sand body, which occasionally even led to sinter formation (Ellenberg & R.-D. Kahlke 1997, 45, pl. 13, 1-3). As a result, the mineral substance of the majority of the

[1] The site survey for the field sheets was carried out by: Dipl.- Geol. G. Braniek, N. Fiebig, Dr. R.-D. Kahlke, J.-A. Keiler, T. Korn, Dr. L. C. Maul, V. Neé, K. Neumann, J. Podschun, U. Rohmer, Dipl.-Phil. U. Steiner, G. v. Stuckrad, and H. Wöllner. From the year 2000 F. Krause, D. Rössler, G. Utschig, M. Stache, M. Stebich, and J. Wagner also participated in the excavations. Exact field sheets as well as the present excavation plans were executed by the draughtsmen H. Wöllner (until 1992) and N. Fiebig (from 1992). J.-A. Keiler and Dr. L. C. Maul partook in the production of the general plans (Figs. 41, 42)

[2] Together with the field sheets and the photographic documentaion of the site, the exact field sheets are filed in the Research Unit for Quaternary Palaeontology, Weimar, of the Senckenberg Society for Nature Research.

[3] In just a few cases, bones indicated by a broken contour line are also found in the northern part of the excavation area (plans VIII, IX, XI, and XII). This concerns specimens from the uppermost level of the excavation. They were preserved in an almost pulverized state and could not be recovered (cf. R.-D. Kahlke 1997b, 282).

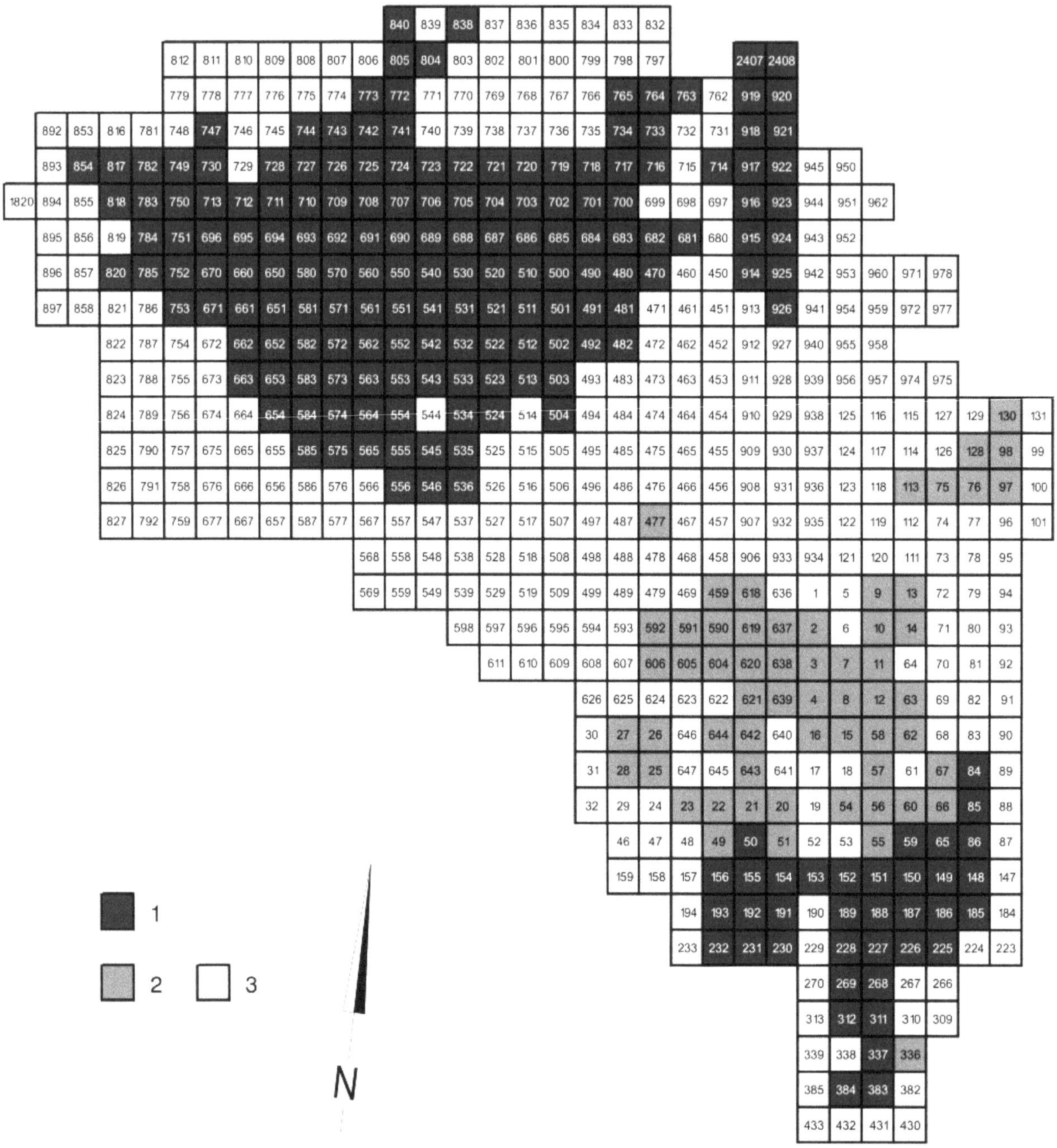

Fig. 41 Research excavation at Untermassfeld. General plan of the excavated area (situation at the end of 1995): 1 Plan squares with fully documented large mammal finds. – 2 Plan squares with incompletely documented large mammal finds, i.e., without fixation of the depths at which they were made. – 3 Plan squares where finds were either not made or lost, e.g., in the central part, where large mammal finds made during the salvage excavation remained undocumented.

vertebrate remains was well preserved. Increased carbonate values in relation to the contribution of phosphates document remineralisation processes within the bones (Keiler 1995, 7). Strongly demineralised specimens, however, which exact a greater effort towards their preservation, were registered from the highly fossiliferous deepest part of the channel (Upper Fluviatile Sands), in and directly above the amalgamation zone of the weathered coarse gravels (cf. Section 2. and R.-D. Kahlke 2001d, 437). Here aggressive water from pores in the secondarily carbonate-free gravels superficially dissolved the fossil material.

In the western to southwestern part of the site (i.e., towards the valley) the fossil bearing Upper Fluviatile Sands (Channel Infill) increasingly dip to the SW. Because it was impermeable to water, the so-called Silt Band, which was formed on top of the highly fossiliferous lower part of the Upper Sands, here kept the fossiliferous stratum from being strongly permeated by moisture and hence from running off to the valley. Its fossil content was

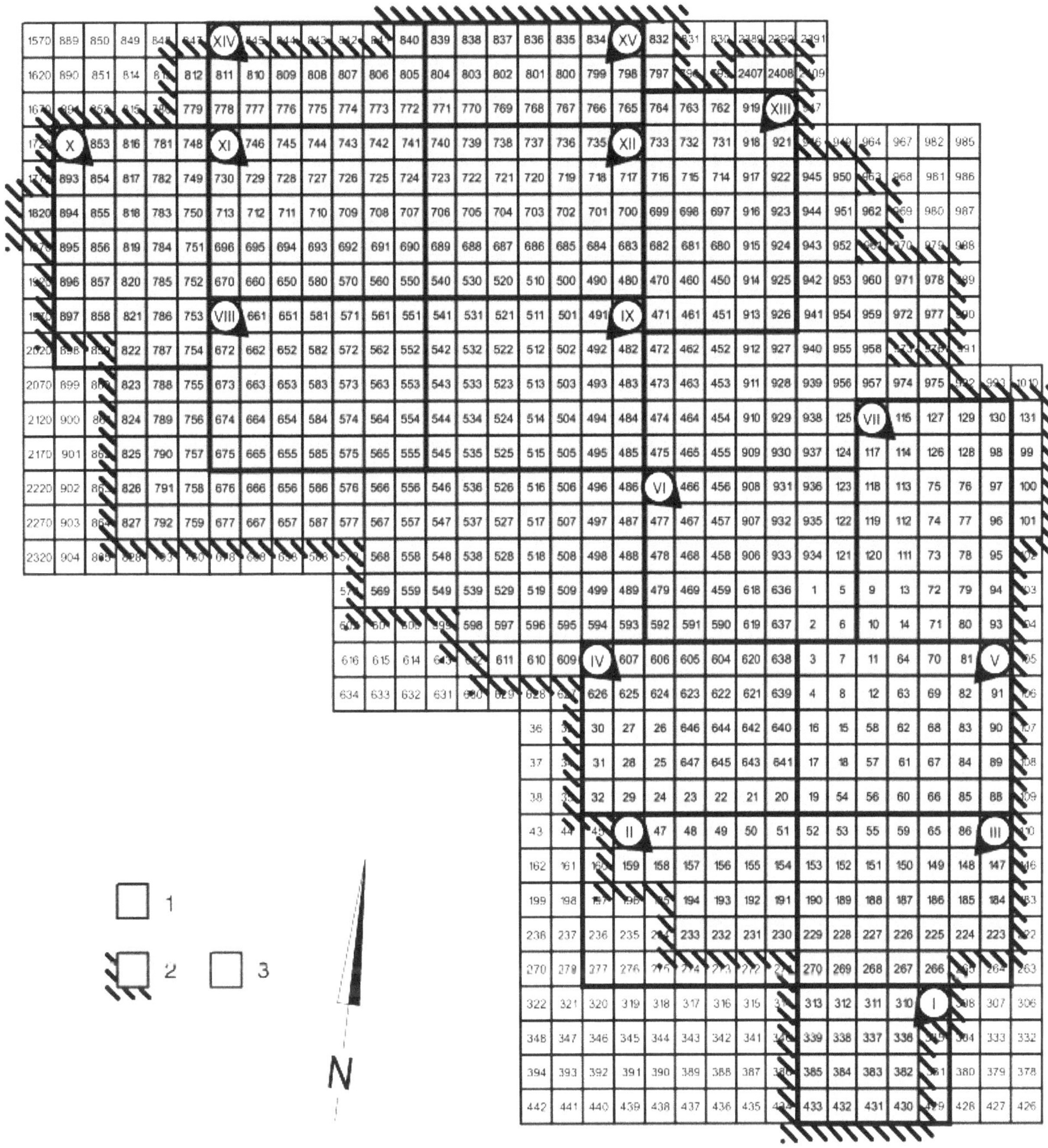

Fig. 42 Research excavation at Untermassfeld. General plan showing the position of the excavation plans I-XV (see supplements): 1 Excavated area (situation at the end of 1995). – 2 Excavated area with outer margin. – 3 Plan squares outside the excavated area.

therefore protected from secondary mechanical damage or destruction (cf. Section 2 and Fig. 3).

Deformed or secondarily broken bones (e.g., a skull of *Hippopotamus*; R.-D. Kahlke 2001e) were obtained from those parts of the sand body that experienced particularly intensive neotectonic stresses (cf. Ellenberg & R.-D. Kahlke 1997). In the middle to northern part of the excavation area, the fossils could have been exposed to additional mechanical loads through movements of heavy quarrying equipment, and sand transport on a track with concrete planking. This was removed as late as 1985 (Ellenberg & R.-D. Kahlke 1997, fig. 9).

Because they initially contain a greater quantity of non-mineral substance, the fossil remains of young animals generally prove to be more porous when compared to specimens that belong to adults. They will therefore more readily suffer from water and frost, which became apparent especially within the uppermost section of the profile. Almost pulverized skeletal elements have now and then been recorded here.

The large majority of the recovered remains of both large mammals and small vertebrates show no signs of rolling caused by transport. Exposed sharp-angled edges of articular surfaces of specimens found isolated had also generally remained undamaged. Just a few much-rolled skeletal elements of large mammals had apparently been transported over greater distances as detached fleshless elements and are exceptions to the rule. In addition to the not unusual discovery of articulated associations of bones (cf. Section 4.4.1.), the almost complete absence of traces of wear and weathering indicates rapid covering of the material. Complete bones often show cavities not filled with sediment. In such cases largely fresh skeletal elements were embedded with intact periosteum and corresponding fat content (cf. also Arribas & Palmqvist 1998, 34). The marks of etching, brought about on the bone surfaces by plant roots, were only produced after the bones had been covered by sediment (cf. Section 5.4.).

4.3. Large-scale spatial distribution of the finds

The distributional pattern of the bone finds from the documented excavation areas (supplements I-XV) first of all mirrors a series of geological characteristics of the site. Especially in the northern to northwestern area of the excavation, the increasing dip of the fossiliferous sands towards the SW becomes apparent in a tendency of the specimens to be found at increasingly lower levels in that direction [compare the find depths on the excavation plans VIII and XI (W) with those on IX and XII (E)]. In addition, large concentrations of finds clearly trace the opened-up section of the deepest part of the channel, which strikes SE/NW (excavation plans VIII, X, XI/ Q 564-783).

4.3.1. Channel Infill

Within the Channel Infill (the lower part of the Upper Fluviatile Sands), which is the main section of the excavation, the density of finds – departing from the northern boundary of the Coarse Clastic Mudflow Fan (Ellenberg & R.-D. Kahlke 1997, fig. 5) – increases in a northerly to northwesterly direction. Although the relations within the undocumented central part of the excavated area remain unclear, the northern sections of the excavation (especially those rendered on excavation plans VIII-XII) show distinctly larger find concentrations than the southernmost ones (excavation plans I-III). To judge by the quantity of finds recovered from the central part undocumented (cf. Section 4.1.), it seems likely that appreciable find concentrations were also encountered in the area 12-20 m north of the Coarse Clastic Mudflow Fan [oral comm. by H.-D. Kahlke and H. Wöllner (Weimar)]. To date, the northern and western limits of the fossil lagerstätte have not been reached by the digging (R.-D. Kahlke 2001c).

In close vicinity to the Coarse Clastic Mudflow Fan, which had been formed south of the area where the finds were made, most specimens were relatively small and light (excavation plans I-III), whereas many heavy skeletal elements or connected clusters of specimens were encountered in the middle part of the excavated area (especially excavation plans IV-VI), for instance several extremities of *Hippopotamus* (cf. also R.-D. Kahlke 1997d, figs. 2-4). A slight trend towards decreasing weight of the discovered skeletal remains is again observable in the northern area of the site (excavation plans VIII-XII). The northwestern section of the deepest part of the channel mostly contained skeletal remains of about average weight (excavation plans VIII, X, XI/ Q 582-783). A femoral diaphysis of *Mammuthus* (Q 784-785), a largely complete pelvis of *Stephanorhinus etruscus* (Q 749-750, Fig. 46a), and a humerus fragment in articulation with a radius/ulna of *Hippopotamus amphibius antiquus* (Q 854) form an exception here (excavation plan X).

The spatial distribution of articulated (primary) and hence in most cases relatively heavy clusters of specimens correlates with the frequency of larger isolated elements. In the southern section of the excavated area, directly north of the Coarse Clastic Mudflow Fan, there are no anatomically connected finds of large mammals whatsoever. On the other hand, the fragmentation of isolated elements has proceeded to a considerable degree here (excavation plans I-III). Probably the highest proportion of clusters of articulated specimens comes from the central part of the site, from both the documented (excavation plans IV-VII) and the undocumented areas. In the northern section of the excavation a slight decrease in the number of articulated clusters becomes apparent (excavation plans IX, XII). These were essentially found outside the deepest part of the channel, where finds were abundant. The basal part of the Channel Infill itself (excavation plans VIII, X, XI/ Q 564-783) contained more than 100 disarticulated, and in some measure fragmented, skeletal elements per m^2 (channel lag *sensu* Behrensmeyer 1988).

Orientation of the long axes of bone clusters or isolated skeletal elements was observed in some squares within the deepest part of the channel, but hardly at all in the remaining part of the Channel Infill. Thus an incipient orientation of long bones parallel to the direction of the current becomes discernible on the plan in square Q 582, occasionally also in the adjacent squares Q 551 and Q 581 (excavation plan VIII), as well as in Q 660 (excavation plan XI). Conversely, the left hind-extremity of a *Hippopotamus* (excavation plan VIII/ Q 564; cf. also R.-D. Kahlke 2001e, fig. 3) and a few long bones (excavation plan X/ Q 783, Q 784-785) were oriented at right angles to the direction of the current. Something similar applies to ribs and an antler beam attached to a skull in Q 713 (excavation plan XI).

In summary, it may be concluded that the finds within the Channel Infill are heterogeneously distributed. Whereas completely disarticulated, mostly smaller, lighter and often already considerably fragmented skeletal elements predominate directly under the lee of the Coarse Clastic Mudflow Fan, the find density, the size / weight, and the degree of articulation of the material increase towards the central area. In the northern to northwestern sections of the examined area, the average weight and the proportion of articulated bone clusters slightly drop again. With uncommonly high concentrations of in most cases fully disarticulated and in part fragmented skeletal elements of mainly average size, the deepest part of the channel occupies a special place. When approaching the northern to northwestern limit of the fossil accumulation in succession, one may expect a gradual decrease of both the average size and the weight per specimen within the entire Channel Infill.

4.3.2. Flooded surfaces

Statements regarding the larger-scale distribution of the bone finds outside the Channel Infill cannot be made with similar exactitude, because within the Lower Fluviatile Sands only a comparatively small area has been excavated and documented (the NE part of excavation plan VII, the E part of excavation plan XII, and excavation plan XIII). A survey indicating the position of the NE flank of the erosional channel (boundary Lower / Upper Fluviatile Sands), at datum level of the site, is to be found in R.-D. Kahlke (2001d, fig. 9).

In the part of the Lower Fluviatile Sands examined until now, the find concentrations encountered are generally smaller than in the sections of the Channel Infill adjacent to the west (cf. especially excavation plans VIII-XII). A noticeable exception, however, becomes apparent at the eastern border of the area excavated within the Lower Fluviatile Sands (excavation plan XIII/ Q 915-918, Q 921-924). With approximately 800 separate finds (digging campaign 1993), the more-or-less N-S-oriented bone concentration that was uncovered here displays an extremely high find density. The skeletal remains were dispersed through several sand layers within a height interval of 1.85 m [0.65 m below (Q 924) to 1.20 m above datum level (Q 922)]. The majority of the finds were recorded between 0 and 1.00 m above datum level (cf. excavation plan XIII). Several largely complete extremities (hind-extremity of *Eucladoceros giulii*, fore-extremity of *Bison menneri*, fore-extremity of *Stephanorhinus etruscus*) or articulated parts of such associations (hind-extremity of *Cervus* s.l. *nestii vallonnetensis*, two fore-extremities and a hind-extremity of *Eucladoceros giulii*), comparatively short portions of the cervical vertebral column of *Eucladoceros giulii* and a partial skeleton of *Canis mosbachensis*, were found integrated in the bone concentration (Q 923-924)[4]. The long axes of nearly all long bones and mandibular rami are evidently oriented more or less parallel to the main current direction, following the probable outline of the riverbank, and create an image that deviates from the other distributional patterns at the site. The said connected limb-bones likewise exhibit the general orientation. In one particular case the flaccid system of sinews was even bent correspondingly (hand skeleton of *Stephanorhinus etruscus* in Q 923).

It is possible that in the squares Q 915-918 and 921-924 one is dealing with a perhaps repeatedly renewed washout seam, formed under a shallow covering of water. Within the excavated area mentioned, the zone directly adjacent to the riverbank was certainly not cut, because the strata so far exposed demonstrate that the individual layers of the Lower Fluviatile Sands neither wedge out nor even thin out at all. Yet the different levels of the respective finds in the various sand layers indicate a largely constant fluviatile subenvironment.

All in all, the bone material from the Lower Fluviatile Sands is largely disarticulated, though seldom fragmented. Among the finds made outside the bone concentration within the plan squares Q 915-924 were only one articulated fore-extremity of *Stephanorhinus etruscus*, rather badly preserved at the side in contact with the overlying sediment (excavation plan XII/ Q 701-702; cf. also Keiler 1995, pl. 4-5; R.-D. Kahlke 2000b), and a skull of *Canis mosbachensis* (excavation plan XII/ Q 700) with five articulated cervical vertebrae (Fig. 35).

[4] On excavation plan XIII these units are only discernable with difficulty. The corresponding exact field sheets show the find situations in a higher graphical resolution.

4.4. Smaller-scale distribution of the finds

An analysis of the smaller-scale distribution of the finds allow detailed statements regarding the order of events within the process of disarticulation and accumulation of the carcasses or skeletal material carried into the site. In essence all bone remains are distributed according to one of the following patterns:

Anatomically connected units
Isolated finds
Polyspecific bone concentrations
Bone accumulations

Complete skeletons of larger mammals are absent.

4.4.1. Anatomically connected units

Anatomically connected units usually are completely or partially preserved extremities, each dislodged from the shoulder or hip girdle (e.g., Figs. 43, 44, and excavation plans IV/ Q 642; V/ Q 58; IV-VI/ Q 605/606, 591/592; VII/ Q 9/10/13/14; VIII/ Q 584; IX/ Q 521; XII/ Q 688, Q 705/706; XII/ Q 720; XIII/ Q 923 with four units). Intact portions of vertebral columns have been observed only rarely (e.g., excavation plans VIII/ Q 585; XII/ Q 520, Q 704, Q 704/705, Q 702, Q 722). The anatomically connected units are characterized by normal relationships between the positions of the skeletal elements, or by such disarticulations as reveal immediately that the isolated specimens belong together. Extreme distortions of articulated units, as they arise from contraction of the sinews through sudden desiccation of carcasses, have not been observed.

Surveys regarding the taphonomical characteristics of the recovered skeletal elements of each of the more frequently encountered large herbivores (*Hippopotamus amphibius antiquus*, *Stephanorhinus etruscus*, *Bison menneri*, *Eucladoceros giulii*, *Equus wuesti*, *Cervus* s.l. *nestii vallonnetensis*) were drawn up (Figs. 49-72). Of these species a total number of 48 anatomically connected units exist, whereas an additional number of 112 units could be reconstructed from isolated specimens (Table 1, Figs. 52, 56, 60, 64, 68, and 72, cf. Section 4.6.).

Furthermore, Sotnikova (2001) described three anatomically connected units that belong to *Canis mosbachensis*: one of the vertebral column, and two of extremities. In 1993 another and hitherto uninvestigated individual unit (partial skeleton) of this small canid was recovered from the plan squares Q 923 and Q 924 (cf. Section 3.2.6.). The wolf skull with articulated cervical vertebrae already mentioned above came from square Q 700 (digging campaign 1994). For *Spermophilus* an anatomically connected unit consisting of an upper and a lower jaw (Q 555) was already documented by Maul (2001a, 860). Meanwhile ten skeletons (Fig. 37) or partial skeletons of small mammals have been found (digging campaigns 2002 and 2003: Q 24, Q 47, Q

	anatomically connected units		reconstructed units		total
	vertebral column	extre-mities	vertebral column	extre-mities	
Hippopotamus amphibius antiquus	3	11	0	5	19
Stephanorhinus etruscus	1	3[5]	1	40	45
Bison menneri	1	9	0	28	38
Eucladoceros giulii	6	6	0	21	33
Equus wuesti	0	7	0	13	20
Cervus s.l. *nestii vallonnetensis*	0	1	0	4	5
sum	11	37	1	111	160

Table 1 Distribution of anatomically connected units and reconstructed units of the more frequently represented large herbivores.

[5] Anatomically connected units recovered at Untermassfeld at the time of the salvage excavations (1979-1982) could with a high degree of probability be reconstructed on the basis of date of recovery and state of preservation. They are included in the analysis below.

Fig. 43 Research excavation at Untermassfeld. Left fore-extremity of *Hippopotamus amphibius antiquus* and parts of the right foot skeleton of *Stephanorhinus etruscus in situ* (plan squares Q 591-592, 605-606, October 1980). Height represented by the photo cutout: 1.12 m.

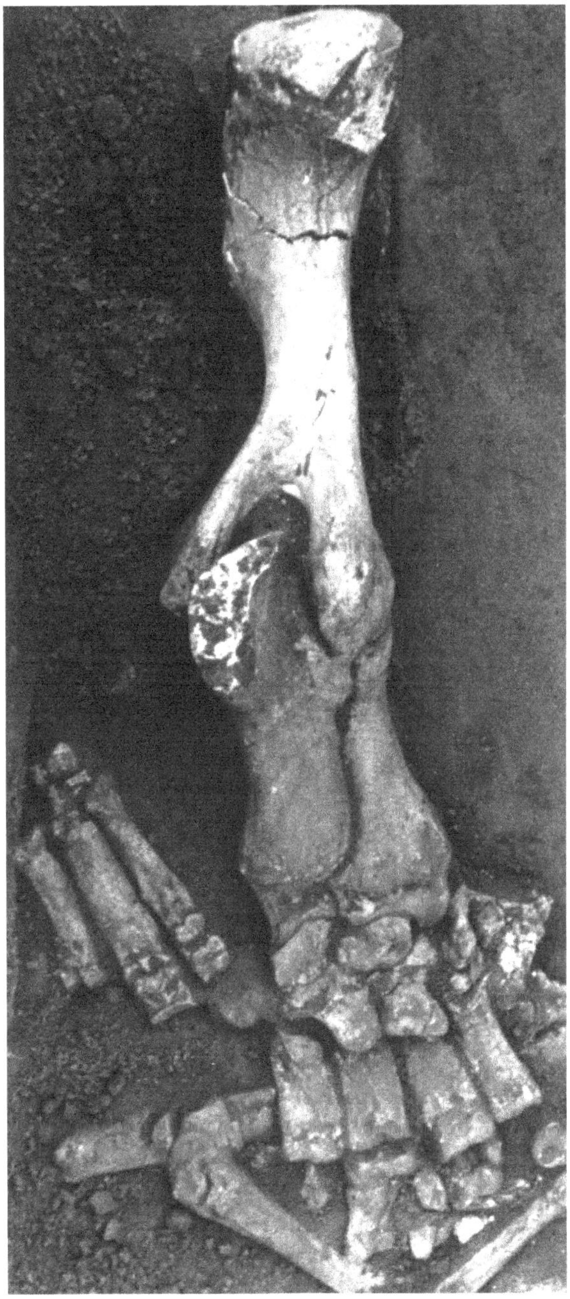

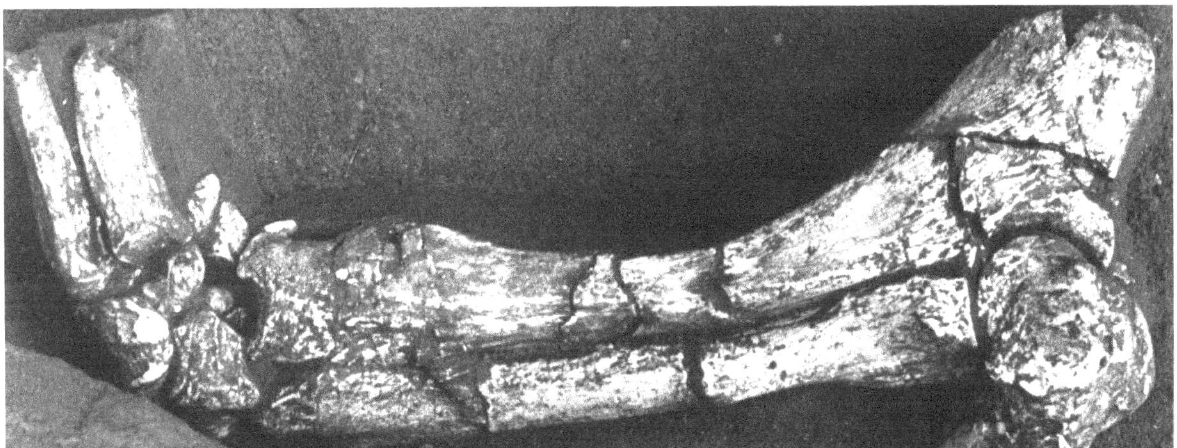

Fig. 44 Research excavation at Untermassfeld. Right forearm and hand skeleton of *Stephanorhinus etruscus in situ* (plan square Q 521, October 1984). Width represented by the photo cutout: 0.71 m.

156-157, Q 193-194). In addition, Jánossy (1997, 98-99) described an individual unit of a raven (*Corvus* sp.)

An exemplary sequence of disarticulation events which remained largely uninfluenced by carnivores was described from the sedimentation area of the present site by means of linking a number of selected find situations involving *Hippopotamus* extremities (R.-D. Kahlke 1997d, figs. 1-6, 1999, figs. 3-8). The reconstructed decomposition process affecting a bear (R.-D. Kahlke & Kierdorf 1997, 162-164) and a *Hippopotamus* skull (R.-D. Kahlke 2001e, 482ff.) likewise illustrates the destruction of skeletons in moving water.

The impact of most carnivorous mammals shown to be present in the thanatocoenosis of Untermassfeld (cf. Section 3.2.6.) furthered and modified the decomposition of the carcasses or skeletal units before and after they had been carried into the site (cf. Section 4.5., 5.).

4.4.2. Isolated finds

Because single elements became detached or because skeletal units of large mammals completely disarticulated, a large quantity of isolated finds came into being in the area of the site. The majority of small vertebrate remains were found isolated (Maul 2001a, 858ff.). Abiotic and biotic processes in many ways modified the bones. Fractures were caused by the motion of water (cf. Section 5.3.), possibly also by falling rocks and trampling, and perhaps occasionally by desiccation through strong irradiation by the sun.

4.4.3. Polyspecific bone concentrations

As a result of fluviatile activity in the sedimentation area of the site, isolated detached specimens were accumulated in piles containing remains of different individuals and species (e.g., Fig. 45). It was often observed that large or obstructive skeletal elements (so-called "obstruction bones") formed mechanical barriers and therefore directly caused the deposition of additional skeletal parts (e.g., Fig. 46a, b and excavation plans IV/ Q 638; VIII/ Q 582; X/ Q 749-750, Q 751; XI/ Q 713, Q 693/710) (general remarks upon this subject matter in, e.g., Kleinschmidt 1965, 168ff.). The bone concentrations now and then appear to be imbricated, with their component parts inclined in a more or less upstream direction (e.g., excavation plans IV-VI/ Q 637/638/3; VIII/ Q 582). In some cases, entire anatomically connected extremities are entangled in the aggregations (e.g., excavation plan VIII/ Q 564, and Fig. 17). Furthermore, an increased proportion of finds belonging to small vertebrates was demonstrated by wet-sieving (Maul 2001a, 858). Irregular distributed, fragmentary, flattened pieces of xylite indicate that parts of plants from the former floodplain vegetation were plaited into the bone concentrations (cf. Section 3.1.). From time to time such vegetation remains certainly also enmeshed and aggregated objects transported by the river.

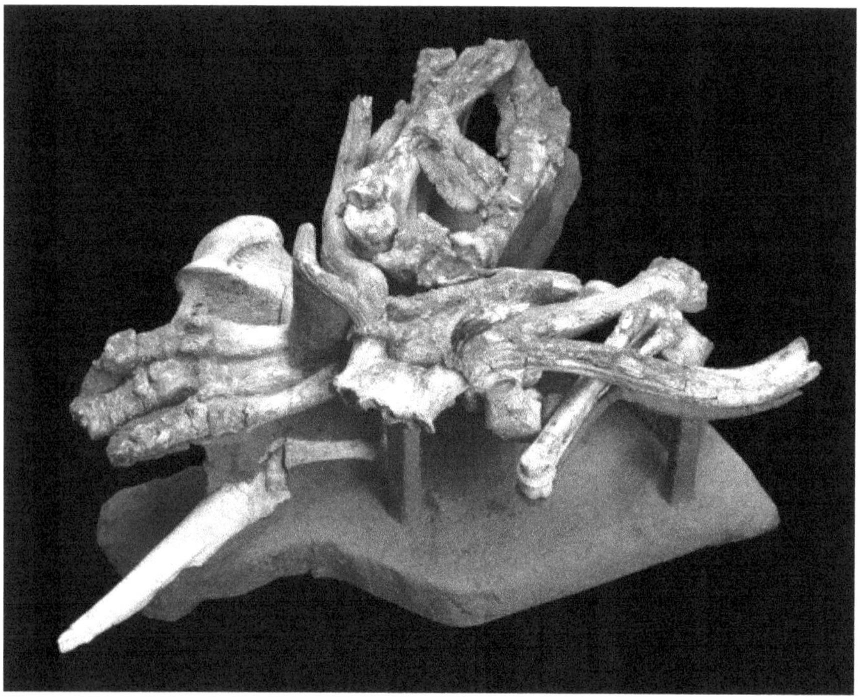

Fig. 45 Polyspecific bone concentration, Untermassfeld. *In-situ* mount with skeletal remains of *Bison menneri*, *Cervus* s.l. *nestii vallonnetensis*, *Eucladoceros giulii*, *Stephanorhinus etruscus*, and *Equus wuesti* (plan squares Q 807-808, Q 842-843) [IQW 2003/ 28 185 (Mei. 27 347)]. Width of the mount: 0.76 m.

4.4.4. Bone accumulations

In the deepest part of the channel in the Upper Fluviatile Sands, a bone accumulation with a constantly high concentration of finds was documented. Approximately 11 m² of its area have thus far been uncovered (e.g., excavation plan XI/ Q 660 to excavation plan X/ Q 817, Q 749) (Fig. 47). As already mentioned, more than 100 disarticulated skeletal elements per m² have been recovered here (on the geological situation regarding the deepest part of the channel, see R.-D. Kahlke 2001d, 437ff.). Such accumulations may be interpreted as a fusion of several polyspecific bone concentrations with a large input of skeletal material. In most cases it is no longer possible to determine the starting point of each accumulation of specimens (the "obstruction bone") with certainty.

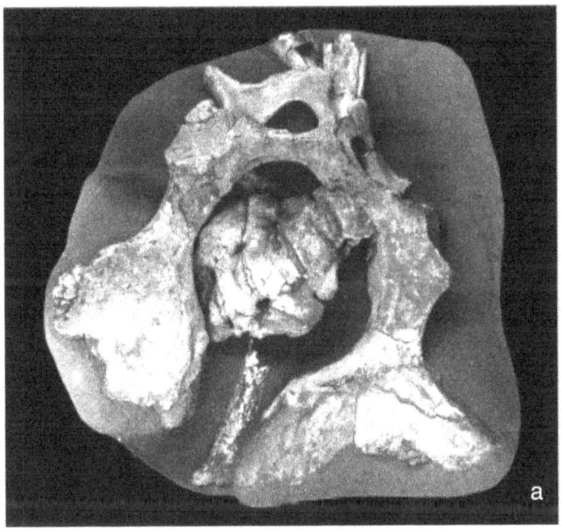
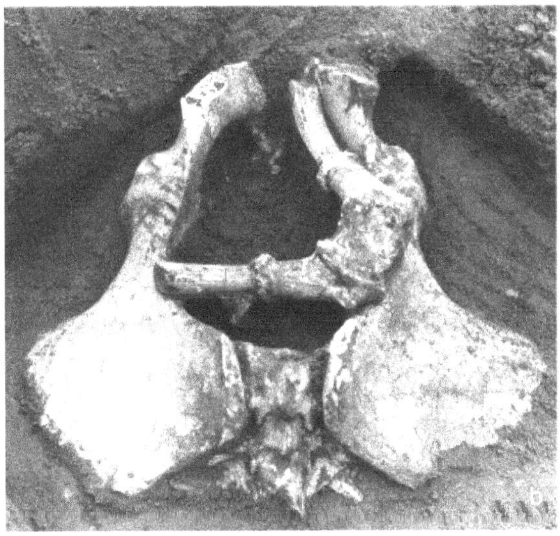

Fig. 46 Research excavation at Untermassfeld. Pelves and sacra of *Stephanorhinus etruscus* as „obstruction bones" that caused additional deposition of skeletal elements: a *In-situ* mount from Q 749-750 [IQW 1997/ 26 334 (Mei. 25 863)]. – b *In situ* at Q 619 [IQW 1980/ 16 113 (Mei. 15 624), IQW 1980/ 16 115A (Mei. 15 626), 1980/ 16 115B (Mei. 15 626), IQW 1980/ 16 435 (Mei. 15 946), IQW 1980/ 16 436 (Mei. 15 947)]. Width represented by the photo cutouts: 0.65 m.

4.5. Trace fossils

Traces of activities by various animal groups may be identified on the recovered material. Data on the distribution of carnivore tooth marks are to be found in the respective contributions on the cervids (H.-D. Kahlke 1997), hippopotamids (R.-D. Kahlke 1997b, 2001e), rhinocerotids (H.-D. Kahlke 2001b), equids (Musil 2001a), ursids (Musil 2001b), hyaenids (A. Turner 2001), and felids (Hemmer 2001), as well as in a separate study on the bifurcations of bone ends (R.-D. Kahlke 2001f). With numerous tooth marks, bone splinters and coprolites, *Pachycrocuta brevirostris* in particular left clear indications of its activities. For example, as many as 11.6 % of the examined skeletal remains of *Hippopotamus amphibius antiquus* show a recurrent type of damage, mainly caused by hyaenas (R.-D. Kahlke 1997b, 279-283). Essentially all categories of bone destruction by *Pachycrocuta brevirostris* – "damage categories" DC 1-9 as they were described from the Early Pleistocene of the Haro River Quarry (NW Pakistan) by Saunders & Dawson (1998, 230ff.) – may be observed on the material from Untermassfeld. A quantitative comparison with the Untermassfeld inventory is desirable for the future.

It is very probable that tooth marks produced by canids with narrower muzzles, similar to those figured by, among others, Miller (1969) and Binford (1981), are also represented at Untermassfeld. It seems unlikely that the gracile teeth of the sabretooth cat *Homotherium crenatidens* and the dirktooth cat *Megantereon cultridens adroveri* made any hard contact with the bones at all. Where the remaining felids are concerned, a forceful penetration of their teeth into the skeletons of their respective prey was presumably rare as well.

Several skeletal elements show sometimes extensive gnaw marks produced by various rodents (Fig. 39) (Maul 2001b). Osteophagia by herbivores is plausible for one metatarsal fragment (R.-D. Kahlke 2001f). Trace fossils of invertebrates are also represented. Coprolites, for instance, presumably show feeding burrows of so far uniden-

Fig. 47 Bone pavement, Untermassfeld. *In-situ* mount with bone concentration from the Lower Fluviatile Sands (deepest part of the channel infill, plan squares Q 48-49) (without number). Width represented by the photo cutout: 0.85m.

tifiable coprophagous insects (Keiler 2001c, 494 f.). In addition, insect damage to bone surfaces is occasionally observed [oral comm. by J.-A. Keiler (Weimar)]. Targetted research will certainly provide more indications of biotic activities.

4.6. Reconstruction of anatomically connected and individual units

With regard to *Hippopotamus amphibius antiquus*, *Stephanorhinus etruscus*, *Bison menneri*, *Eucladoceros giulii*, *Equus wuesti*, and *Cervus* s.l. *nestii vallonnetensis,* the systematic fitting of disarticulated skeletal elements, found isolated or in bone concentrations, led to the reconstruction of a total of 111 – in most cases not quite complete – anatomically connected extremities and of part of a vertebral column (Table 1, Figs. 51, 55, 59, 63, 67, 71).

For the rhinocerotids the completion of "individuals" from several anatomically connected units was repeatedly achieved (H.-D. Kahlke 2001b). This was made possible by a sufficient number of specimens showing the animals' individually variable characteristics – especially body size and age at death – and their representation by a manageable number of finds (651 specimens). From a minimum number of individuals of 29, a total of 12 individuals were represented by substantial portions of the skeleton (4 juveniles, 8 adults). Another individual unit is the partial skeleton of a bear (*Ursus* cf. *dolinensis*) published by R.-D. Kahlke & Kierdorf (1997). Furthermore, Hemmer (2001) identified one individual unit for each of the species *Acinonyx pardinensis pleistocaenicus* and *Puma pardoides*. Additional individual units are represented by the (partial) skeletons of small mammals recovered in 2002, as well as by the partial skeleton of a cheetah from the 2003 campaign (cf. Section 3.2.6.).

The reconstruction of anatomically connected units has certainly not made use of the entire body of material from the site. Both through initial losses in the field (before the establishment of the site) and because the excavation of the bone

	NISP	%
Isolated finds	134	41.9
Indubitable and probable anatomically connected units	160	50.0
Reconstructed units	26	8.1
Total quantity	320	100.0

Table 2 Distribution of the examined skeletal elements of *Hippopotamus amphibius antiquus* with regard to anatomically isolated single finds and skeletal units. NISP: Number of Identifiable Specimens.

	NISP	%
Isolated finds	310	47.6
Indubitable and probable anatomically connected units	56	8.6
Reconstructed units	285	43.8
Total quantity	651	100.0

Table 3 Distribution of the examined skeletal elements of *Stephanorhinus etruscus* with regard to anatomically isolated single finds and skeletal units[6].

	NISP	%
Isolated finds	639	65.5
Indubitable and probable anatomically connected units	83	8.5
Reconstructed units	254	26.0
Total quantity	976	100.0

Table 4 Distribution of the examined skeletal elements of *Bison menneri* with regard to anatomically isolated single finds and skeletal units.

	NISP	%
Isolated finds	718	75.3
Indubitable and probable anatomically connected units	102	10.7
Reconstructed units	134	14.0
Total quantity	954	100.0

Table 5 Distribution of the examined skeletal elements of *Eucladoceros giulii* with regard to anatomically isolated single finds and skeletal units.

	NISP	%
Isolated finds	56	33.5
Indubitable and probable anatomically connected units	39	23.4
Reconstructed units	72	43.1
Total quantity	167	100.0

Table 6 Distribution of the examined skeletal elements of *Equus wuesti* with regard to anatomically isolated single finds and skeletal units.

6 These figures differ somewhat from those presented in H.-D. Kahlke (2001b), because some not straightforwardly articulated single elements can additionally be assigned to the reconstructed individual units published there.

	NISP	%
Isolated finds	301	91.0
Indubitable and probable anatomically connected units	10	3.0
Reconstructed units	20	6.0
Total quantity	331	100.0

Table 7 Distribution of the examined skeletal elements of *Cervus* s.l. *nestii vallonnetensis* with regard to anatomically isolated single finds and skeletal units.

deposit remains unfinished, the total amount of fossils is only partially available. Moreover, it seems likely that not all of the actually existing associations have been recognized. Both the selection caused by fluviatile transport of the (partial) carcasses assembled in the area of the site (cf. Section 4.3.), and carnivores dragging these off afterwards (cf. Sections 5.4., 6.1.), reduced the percentage of reconstructable anatomically connected units. Hence a rather large quantity of represented in the material recovered.

The entire volume of 3399 examined bone specimens of the more abundant large herbivores includes a total of 160 primary or reconstructed anatomically connected units (Table 1). For each of the species considered, the proportions of the skeletal elements occurring within anatomically connected units as rendered in Table 8 follow from the compilations in Tables 2-7 (Fig. 48).

The graphic surveys of Figs. 49-72 document the existing bone spectra of the six large mammal species represented most frequently in the material recovered. For each of the species *Hippopotamus amphibius antiquus*, *Stephanorhinus etruscus*[7], *Bison menneri*, *Eucladoceros giulii*, *Equus wuesti*, and *Cervus* s.l. *nestii vallonnetensis,* the overall distribution of the skeletal elements recovered, and the distribution of isolated elements and of anatomically connected units

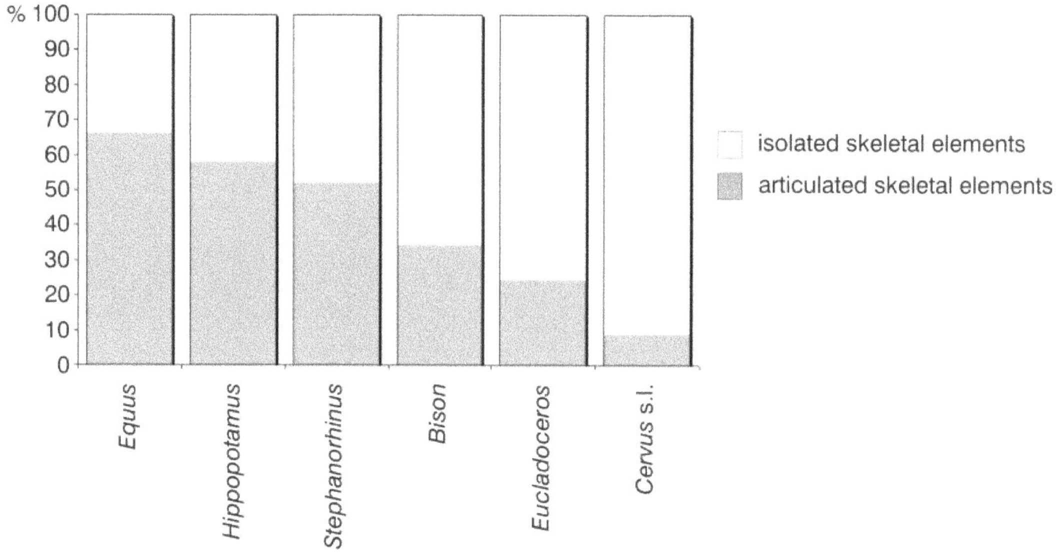

Fig. 48 Untermassfeld. The composition of the recovered material of selected large mammal species represented in terms of percentages.

isolated single specimens remains (Tables 2-7, Figs. 50, 54, 58, 62, 66, 70).

Nearly all species of larger mammals at Untermassfeld have produced dental or skeletal elements from the left and the right body side of the same individuals. *Capreolus cusanoides*, *Megantereon cultridens adroveri*, and *Trogontherium cuvieri* are currently the only exceptions. Each of these is relatively seldom

[7] The skeletal proportions of Figs. 53-56 are based on the dimensions of the largest individual of *Stephanorhinus etruscus* found at Untermassfeld. The graphic design was produced in collaboration with H.-D. Kahlke.

and reconstructed units, are reported in order of decreasing body weight.

Apart from the exceptionally good quality of the material preservation of nearly all finds as they were embedded, it is also the relatively high proportion of anatomically connected and reconstructed units which indicates that the skeletal material was carried into the area of the site within (partial) carcasses. The analyzed finds of faunal elements with various body volumes show that species with a robust build led to clearly increased proportions of individually associated skeletal elements. More gracile forms were more strongly exposed to the manifold biotic and abiotic decomposition processes, both as they were transported into and deposited within the area of the site.

Within the total body of the material examined, the respective proportions of anatomically connected and reconstructed units become smaller with decreasing body size, from *Hippopotamus* via *Stephanorhinus*, *Bison*, and *Eucladoceros* down to the small deer (Fig. 48). Only *Equus* deviates from this regularity, since 66,5% of the material belongs to units that were either found as such or reconstructed. With 167 separate specimens the number of finds is certainly comparatively small, and within this volume a group of individuals that were to a relatively lesser degree exposed to decomposition is overrepresented. The place where they died was probably not very far from that where they were later embedded.

In general the example of the lagerstätte of Untermassfeld, with its find concentrations and large excavated area, shows that the "natural" character of bone accumulations becomes visible in the negative correlation between skeletal robustness and degree of disarticulation if a sufficiently large number of finds is analysed (at least 300-600 specimens per large mammal species) (R.-D. Kahlke 1999). Something similar probably applies to the degree to which skeletal elements of the species involved were destroyed (cf. the case study in Horwitz 1998). Anthropogenic influence would shape the taphonomic ensemble in a more differentiated way, consistent with the subsistence strategy of the moment, viz., through preferences affecting the influx or efflux of bone material and through selective strategies of decomposition and destruction.

4.7. Bone spectra

Within the entire volume of large mammal material from Untermassfeld, complete or near-complete trunk skeletons are not represented at all. Even intact vertebrae are underrepresented in the respective bone spectra. Complete ribs of all species present are counted among the rarer finds. Conversely, their fragments are encountered frequently. The trunk skeletons of the large mammals were therefore subjected to the most forceful destruction. Beside the swelling of the carcasses through gases and the resulting instability of, especially, the thoracic and abdominal cavities, the activities of carnivores concentrated around the carcass were presumably the main cause of this phenomenon. The relatively thin skin cover of the abdomen (but also that of the anal region) often presents a weak spot suitable for tearing a carcass open.

	NISP	%
Equus wuesti	167	66.5
Hippopotamus amphibius antiquus	320	58.1
Stephanorhinus etruscus	651	52.4
Bison menneri	976	34.5
Eucladoceros giulii	954	24.7
Cervus s.l. *nestii vallonnetensis*	331	9.0

Table 8 Proportions of the finds that belong to anatomically connected units of selected large mammal species.

Skulls likewise remained intact only occasionally. They are mostly more or less fragmented, so that the figures quoted in the survey diagrams (Figs. 49-72) are as a rule based on fragments. Obstructive skeletal elements, e.g., complete antlers attached to skulls, or shed beams; bovid skulls with horn cores that jut out; tusks or mandibles of elephants; and complete hippopotamus mandibles with exposed canines and incisives, remained rare as well, or are completely absent. Obviously, such parts of carcasses or skeletons had already become entangled upstream in comparatively large numbers, and probably sometimes got detached from the carcasses and were embedded or destroyed outside the limits of the present site.

Relatively small bones, for example sesamoid bones and ungual phalanges, are likewise clearly underrepresented in the total volume of material. Here, too, some detachment in transport is observable. In keeping with a small yet sufficient transport energy of the current near the bank, such elements from the skeletal units disintegrating within the area of the site were probably carried away in significant quantities.

Species	MNI juveniles / subadults[8]	MNI adults	MNI total
Bison menneri[9]	10+3	27+6	46
Capreolus cusanoides	0	5	5
Alces carnutorum	1	3	4
Cervus s.l. *nestii vallonnetensis*	7	40	47
Eucladoceros giulii	28	32	60
Hippopotamus amphibius antiquus	11	9	20
Sus scrofa priscus	3	3	6
Stephanorhinus etruscus	13	16	29
Equus wuesti	5	8	13
Mammuthus meridionalis ssp.	4	2	6
Panthera onca gombaszoegensis	2	3	5
Acinonyx pardinensis pleistocaenicus	0	2	2
Puma pardoides	0	1	1
Lynx issiodorensis	0	2	2
Megantereon cultridens adroveri	1	1	2
Homotherium crenatidens	0	3	3
Pachycrocuta brevirostris	17	5	22
Meles hollitzeri	0	1	1
Ursus cf. *dolinensis*	2	7	9
Canis (Xenocyon) lycaonoides	0	5	5
Canis lupus mosbachensis	2	16	18
Castor fiber	0	2	2
Trogontherium cuvieri	1	1	2
Hystrix sp.	0	1	1
Macaca sylvanus	0	2	2
Total MNI	110	203	313

Table 9 Summary of individual age composition of larger-sized mammals from Untermassfeld. MNI: Minimum number of individuals.

[8] The respective researchers were generally able to determine the age at death of individuals more accurately than Table 9 suggests.

[9] The number of additional individuals is a conservative estimate based on incompletely prepared new specimens found after the analysis of Sher (1997).

4.8. Minimum numbers of individuals

For the larger mammals examined, the minimum numbers of individuals (MNI), based on quantity and individual characteristics of the recovered teeth and bones, are quoted in Table 9. It shows that the fossil material comprises remains of at least 313 individuals of larger mammals (Fig. 73).

4.9. Mortality patterns

The mortality patterns observed within the available fossil material of the separate "populations" show high proportions of juvenile to young adult animals for most of the more frequently demonstrated species (*Bison menneri*, *Eucladoceros giulii*, *Hippopotamus amphibius antiquus*, *Sus scrofa priscus*, *Stephanorhinus etruscus*, *Equus wuesti*, *Mammuthus*, *Panthera onca gombaszoegensis*, *Pachycrocuta brevirostris*). However, in contrast to the findings concerning the Early Pleistocene assemblage from Venta Micena, Spain (Martinez Navarro & Palmqvist 1999, 58ff.), the quantitative relation between juveniles and adults cannot be explained solely as a result of the carnivores' growing preference for juveniles as the body mass of the prey species increases (cf. Section 5.2.).

By demonstrating the presence of at least ten juveniles (without unequivocally determinable sex), 18 cows, and nine bulls of *Bison menneri*, Sher (1997, 104) was able to draw the image of an age and sex distribution which is conceivable for a natural population. The *Bison* specimens that have been recovered since then (footnote 9) do not contradict this.

In comparison to the other cervids, *Capreolus* and *Alces* each has a low MNI, but a juvenile animal was recorded for the elk, too. On the basis of the wear stages shown by mandibular tooth rows belonging to the available material of *Eucladoceros giulii*, H.-D. Kahlke (1997, 241-243) reconstructed the respective proportions of the age classes. The resulting mortality patterns (Fig. 74) display a high number of juvenile individuals in their first year as well as old individuals aged between 8 and 11 years. Animals in their prime are underrepresented. These patterns have to be interpreted as reflecting a classical U-shaped attritional mortality (principal discussion in R.-D. Kahlke & Gaudzinski 2005). *E. giulii* appears to be a large cervid, exceeding extant European red deer (*Cervus elaphus*) in size and being comparable with the large American elk (*Cervus canadensis*). The estimate of body mass of adult Untermassfeld *Eucladoceros*, calculated from the lengths of the lower molar rows (measurements in H.-D. Kahlke 1997, table 59) by using the allometry formula for cervids of Janis (2003, table 13.3), ranges from 240 to 360 kg (rounded off to the nearest 10 kg), average 276 kg (Fig. 75).

On the other hand, an individual age distribution that markedly differs from that of *Eucladoceros* is documented by the dental remains of the "small deer" *Cervus* s.l. *nestii vallonnetensis* (H.-D. Kahlke 1997, 203-205; principal discussion in R.-D. Kahlke & Gaudzinski 2005). In this case the large majority of finds relate to individuals in the second to fifth year of life (Fig. 76). Body weight estimation for adults of *Cervus* s.l. *nestii vallonnetensis* (measurements in H.-D. Kahlke 1997, tables 22 and 23) resulted in a body mass range between 70 and 130 kg, with an average of 95 kg (Fig. 75). Hence it was similar to extant Mesopotamian (=Persian) fallow deer (*Dama mesopotamica*) and exceeded European *Dama dama* in weight. The different mortality patterns documented for *Cervus* s.l. *nestii vallonnetensis* and *Eucladoceros giulii* can be explained by the different stamina of small or large deer to survive flooding, which are related to their body size and body mass (cf. Section 5.2.).

The total number of individuals of *Hippopotamus amphibius antiquus* comprises 55% juveniles, based on their milk dentition. According to Laws (1968, 46), 5.9-13.8% calves up to an age of one year were observed in larger recent populations. Under normal present-day conditions the mortality rate during the first year of life lies between 45 and 54%.

The suid remains from Untermassfeld belong to at least three juveniles and three fully adult animals, but do not allow more exact conclusions with regard to population dynamics. In the cases of both *Stephanorhinus etruscus* (44.8%) and *Equus wuesti* (38.5%), the contribution of juveniles to the established minimum number of individuals is rather high. Where *Mammuthus* is concerned, 50% of the animals were found to be juveniles or young adults.

It is true that in the entire body of material from Untermassfeld at least 15 individuals of the generally rare felids may be demonstrated to date, but the finds are distributed over six species, so that an interpretation of the age profile has to be omitted. In the case of *Pachycrocuta brevirostris* the situation is more interesting. In contrast to all other carnivores, and also to herbivores, young animals are predominant with 77.3%. On the basis of dental development, A. Turner (2001, 686f.) distinguishes five juvenile age classes, all of which relate to ages below one year. Among the finds from Untermassfeld, *Ursus* cf. *dolinensis*, *Canis (Xenocyon) lycaonoides* as well as *Canis mosbachensis* have so far almost exclusively been represented by adult individuals.

Badger, beaver, and macaques left evidence of only a few individuals.

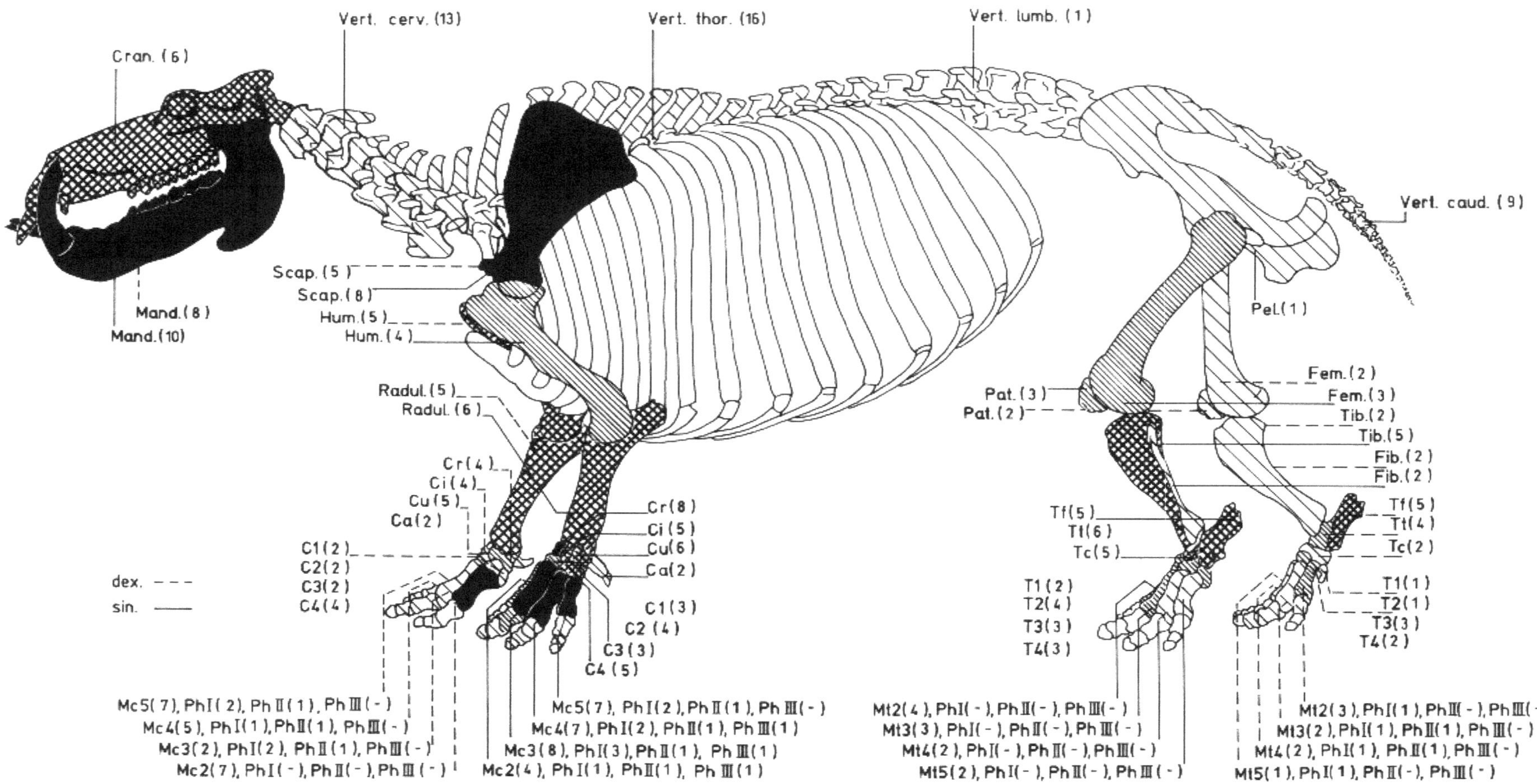

Fig. 49 *Hippopotamus amphibius antiquus*, Untermassfeld.
Categorization of all skeletal elements recovered. In brackets: number of respective specimens.

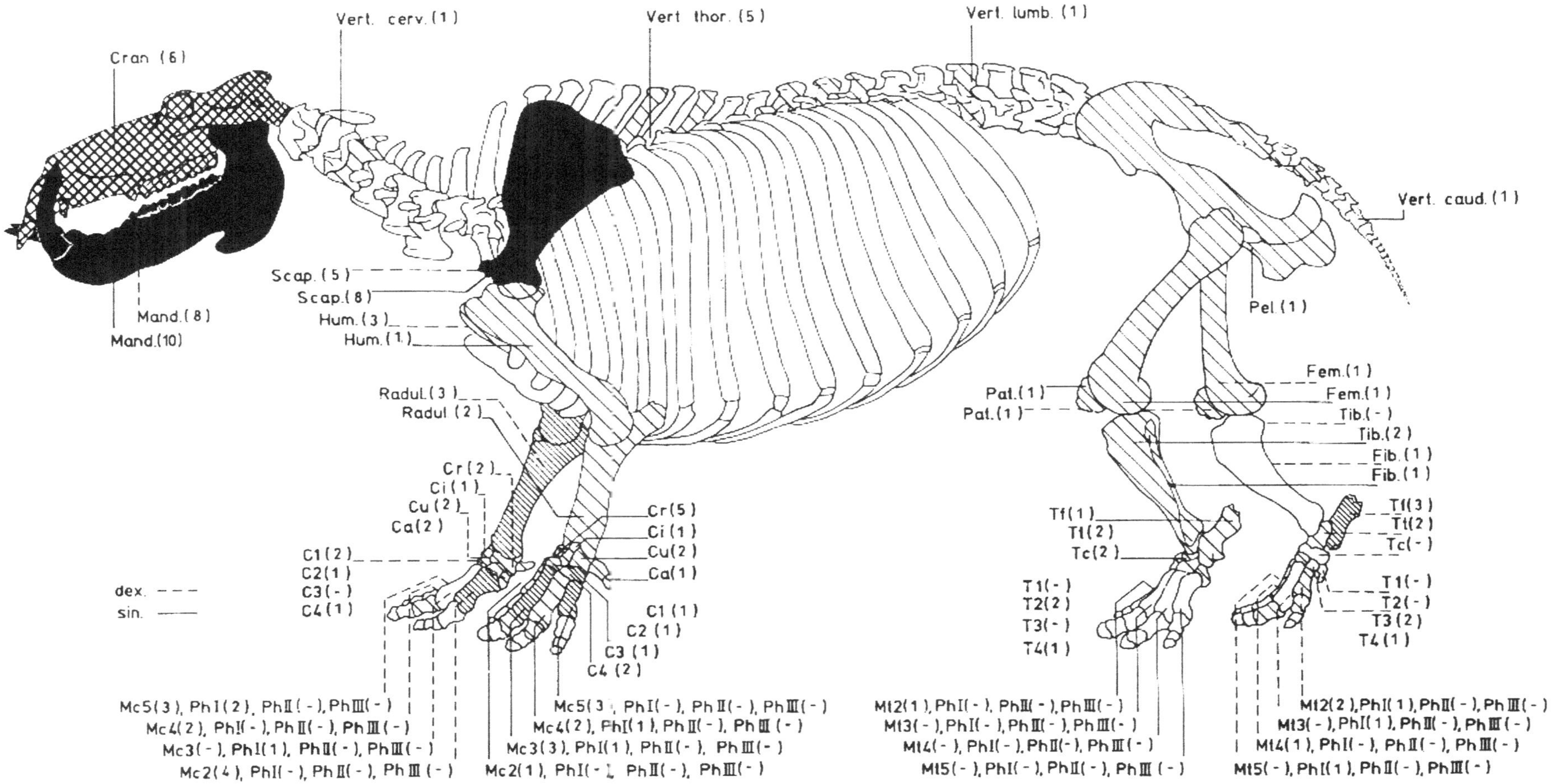

Fig. 50 *Hippopotamus amphibius antiquus*, Untermassfeld.
Categorization of isolated skeletal elements. In brackets: number of respective specimens.

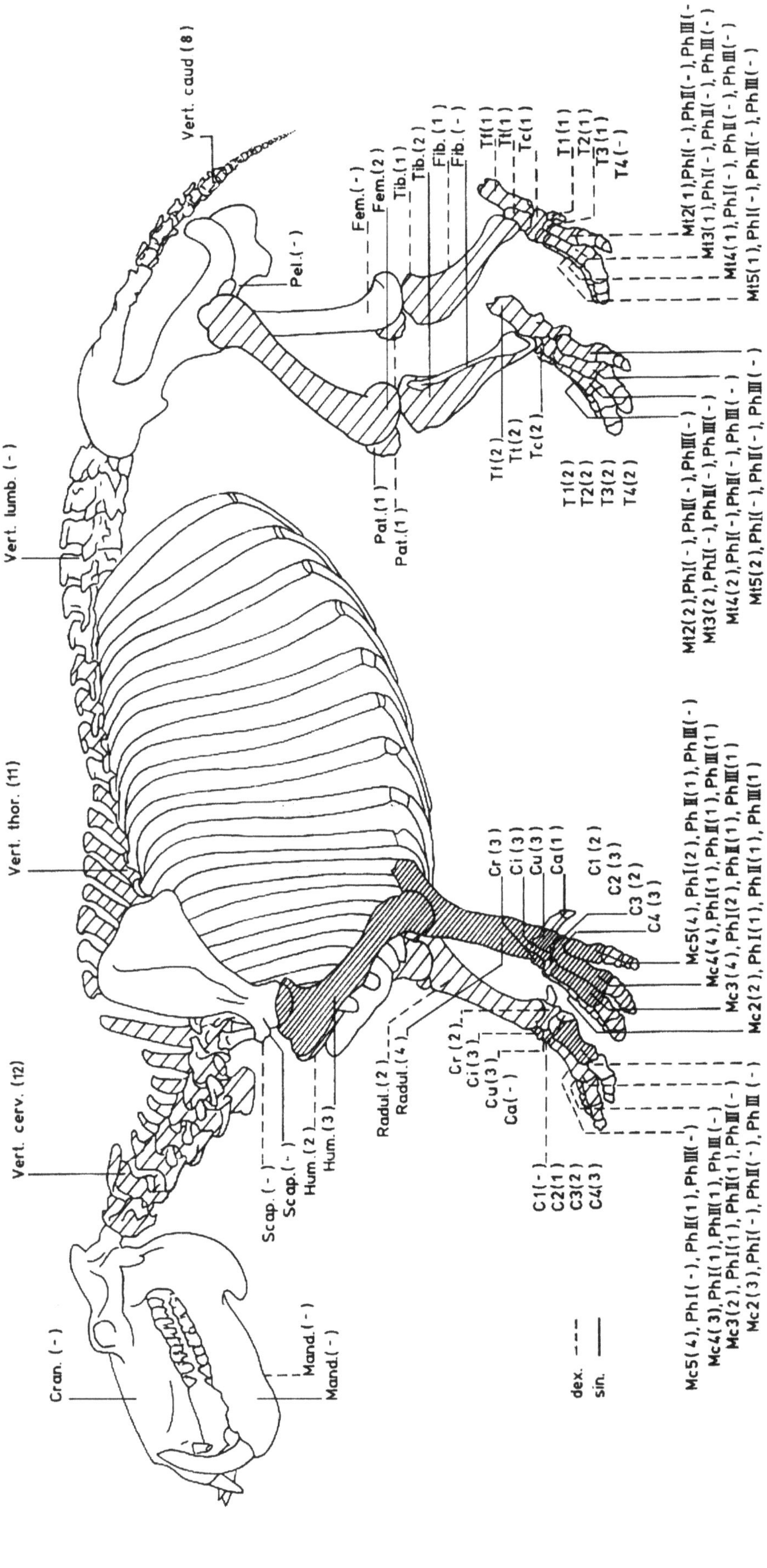

Fig. 51 *Hippopotamus amphibius antiquus*, Untermassfeld.
Categorization of anatomically connected units, and supposed anatomically connected units. In brackets: number of respective specimens.

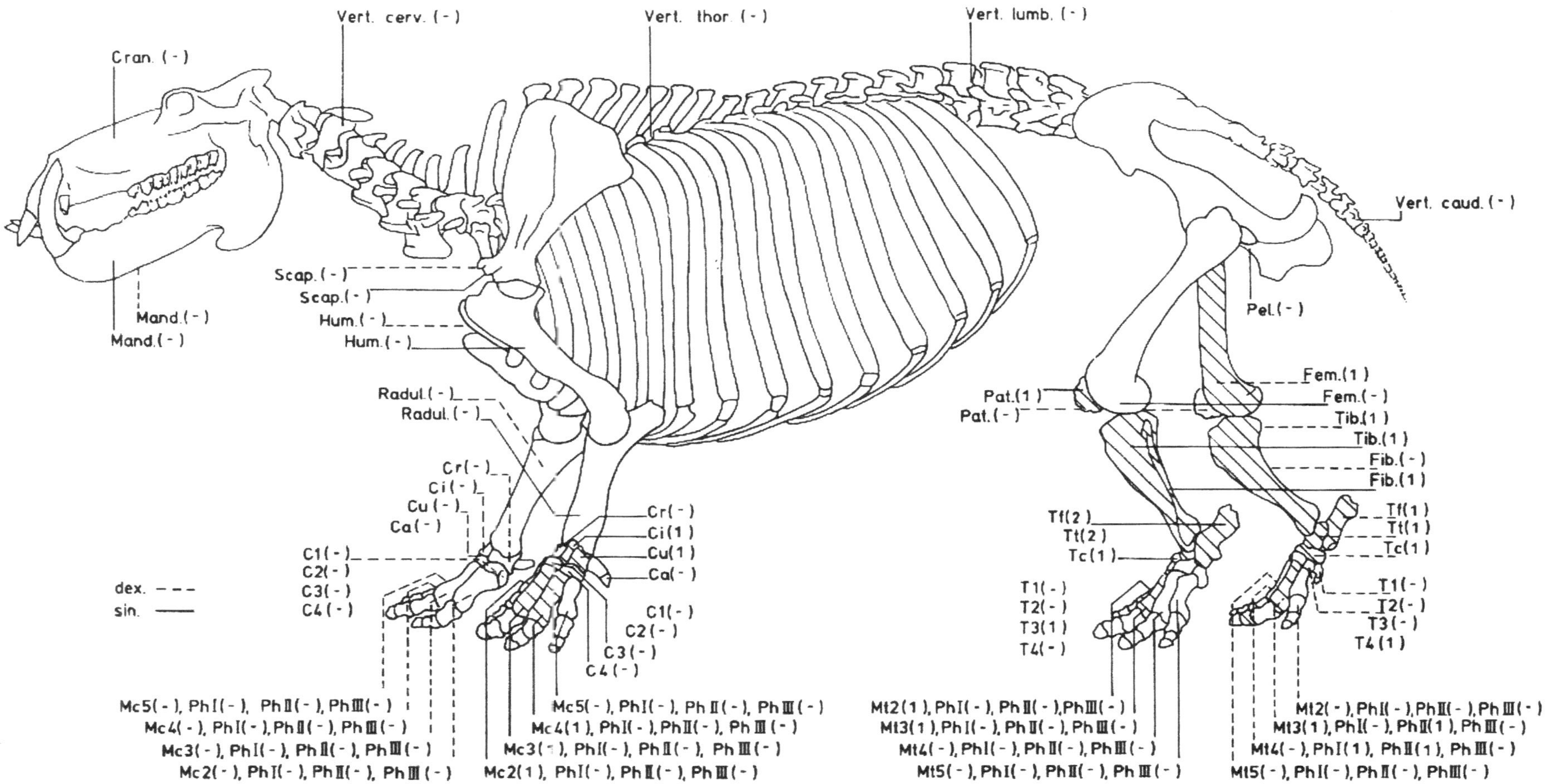

Fig. 52 *Hippopotamus amphibius antiquus*, Untermassfeld.
Categorization of reconstructed anatomically connected units. In brackets: number of respective specimens.

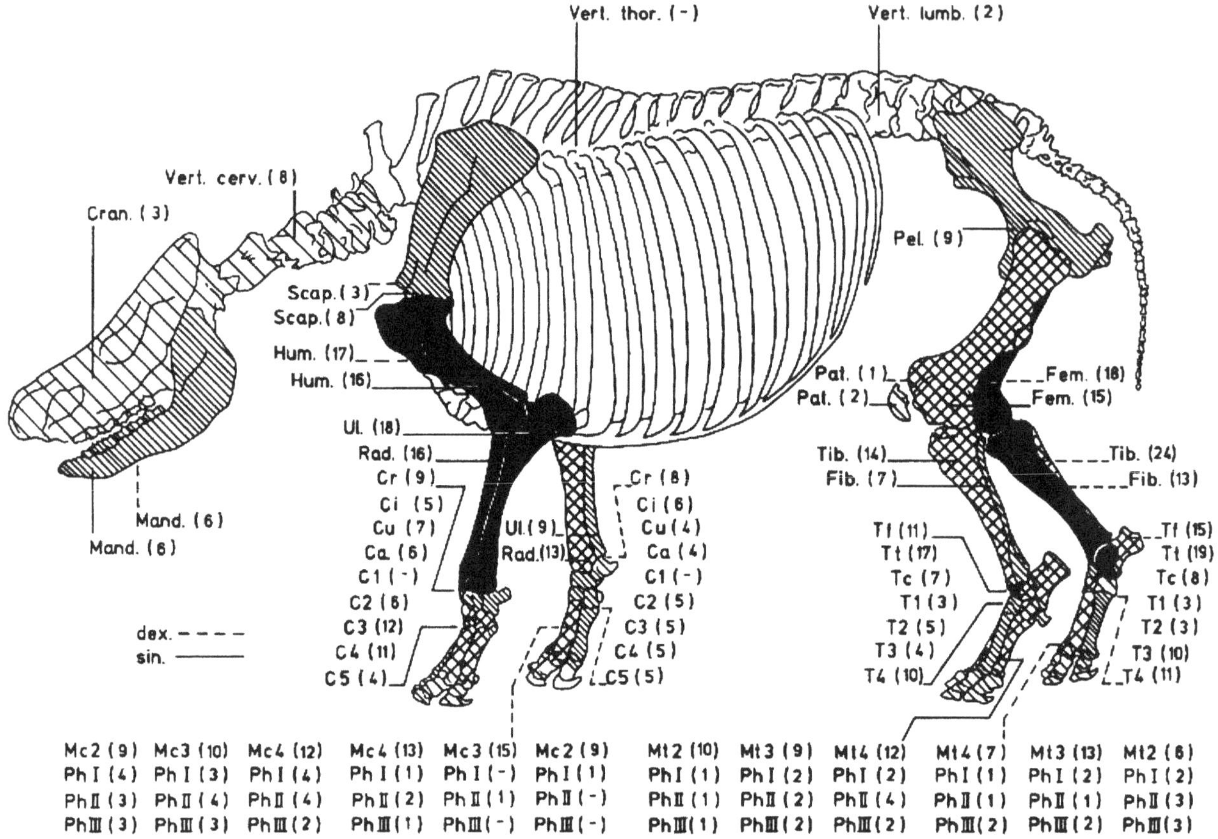

Fig. 53 *Stephanorhinus etruscus*, Untermassfeld.
Categorization of all skeletal elements recovered. In brackets: number of respective specimens.

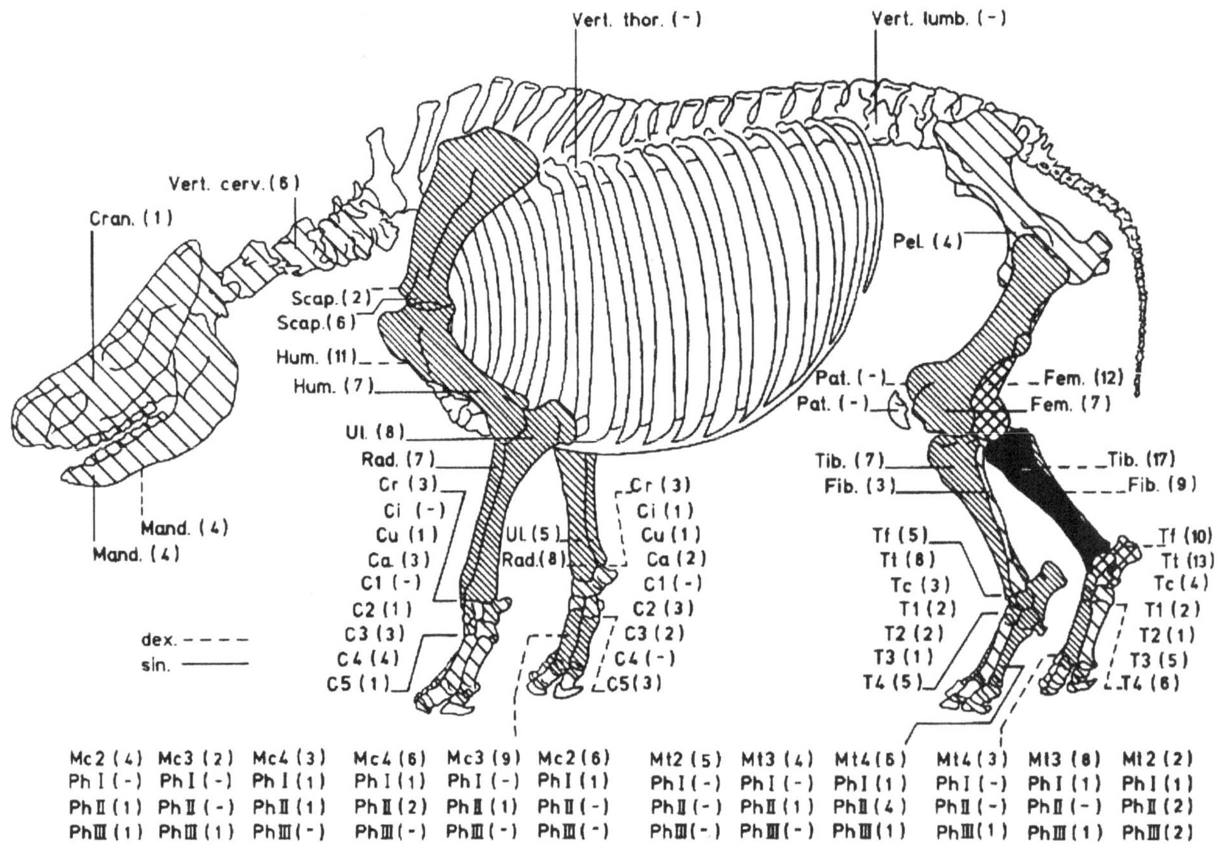

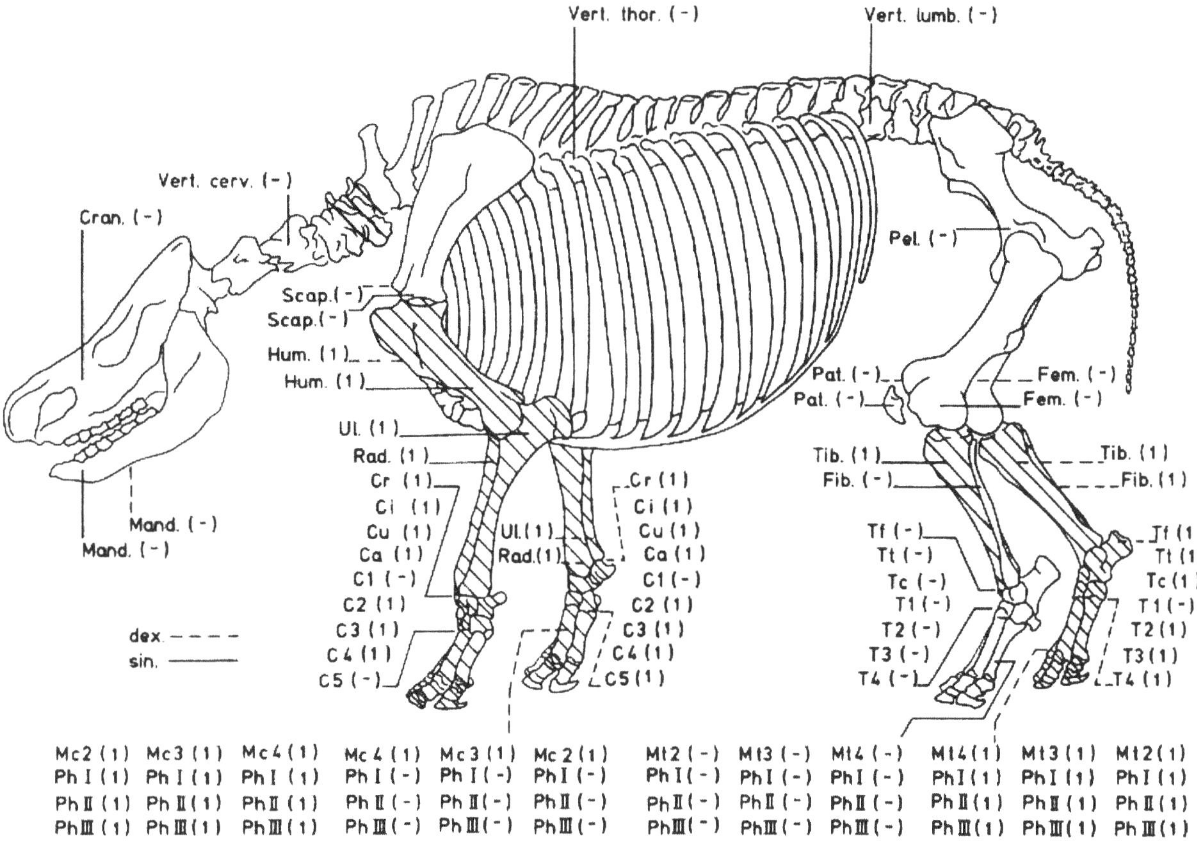

Fig. 55 *Stephanorhinus etruscus*, Untermassfeld.
Categorization of anatomically connected units, and supposed anatomically connected units. In brackets: number of respective specimens.

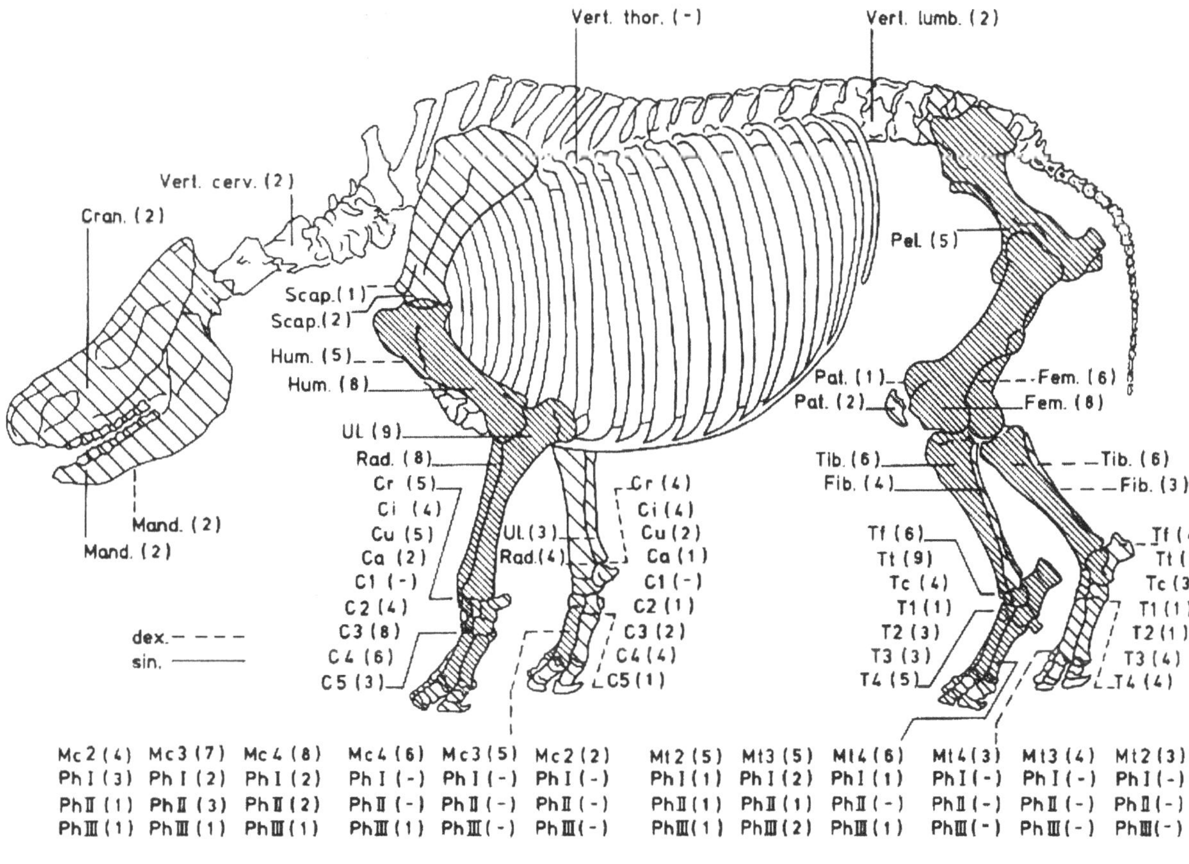

Fig. 56 *Stephanorhinus etruscus*, Untermassfeld.
Categorization of reconstructed anatomically connected units. In brackets: number of respective specimens.

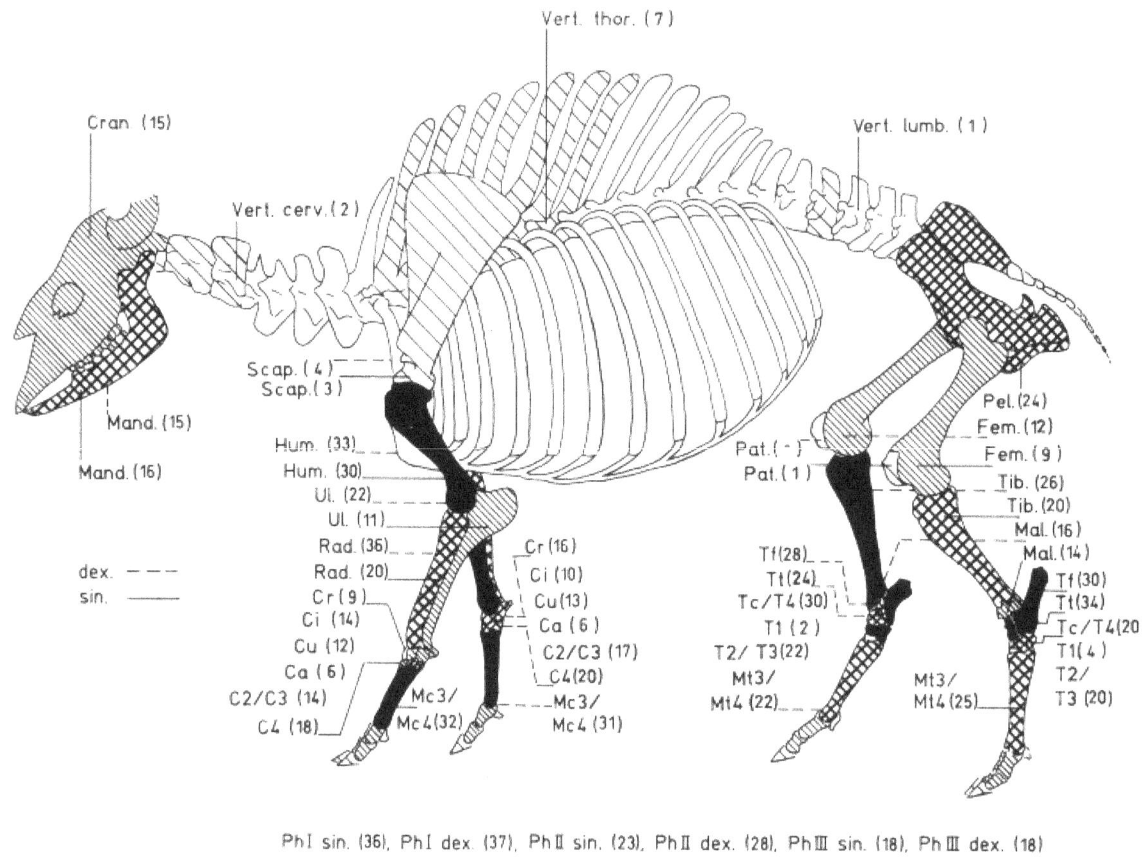

Fig. 57 *Bison menneri*, Untermassfeld.
Categorization of all skeletal elements recovered. In brackets: number of respective specimens.

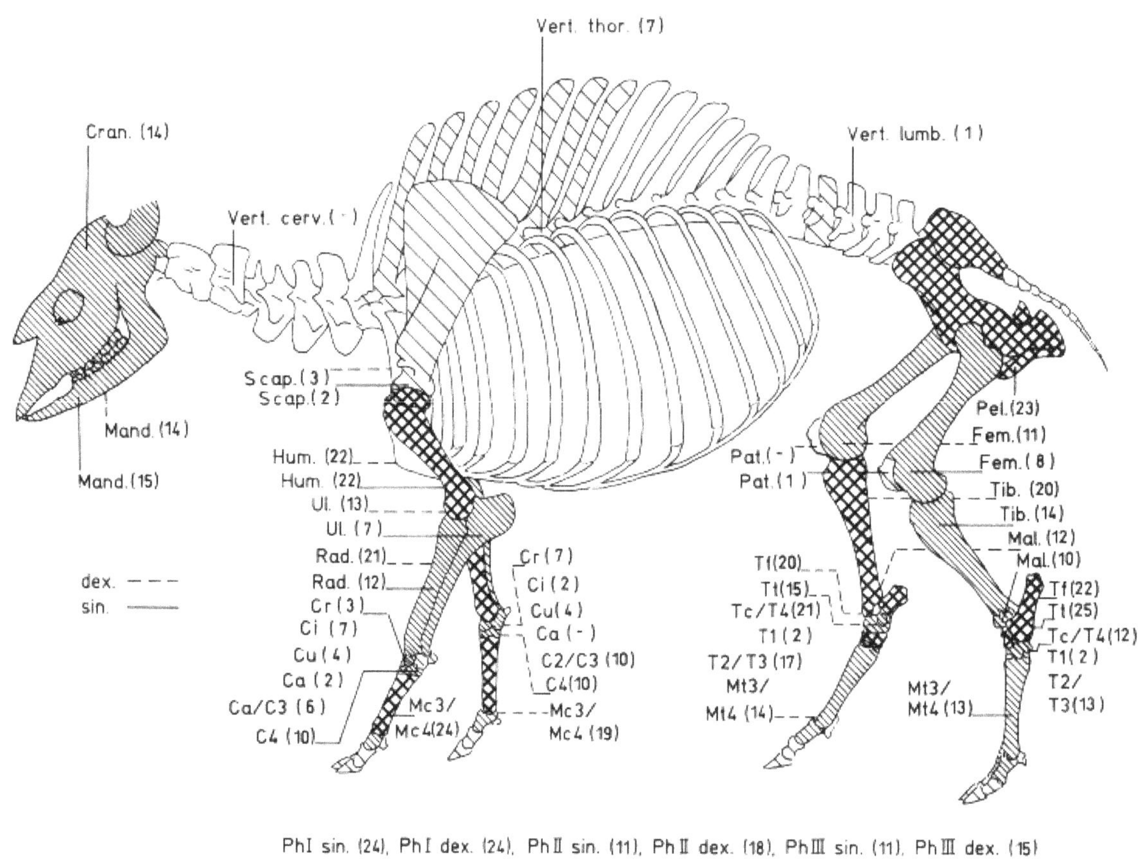

Fig. 58 *Bison menneri*, Untermassfeld.
Categorization of isolated skeletal elements. In brackets: number of respective specimens.

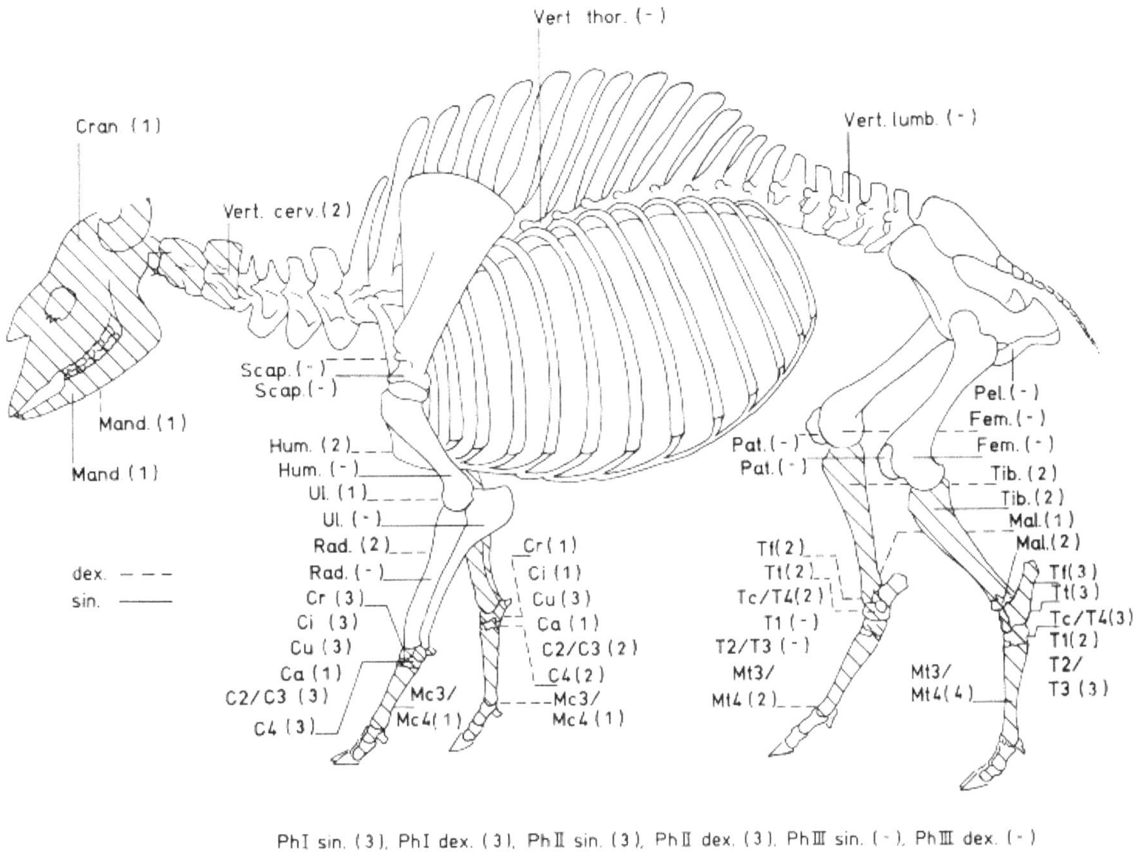

Fig. 59 *Bison menneri*, Untermassfeld.
Categorization of anatomically connected units, and supposed anatomically connected units. In brackets: number of respective specimens.

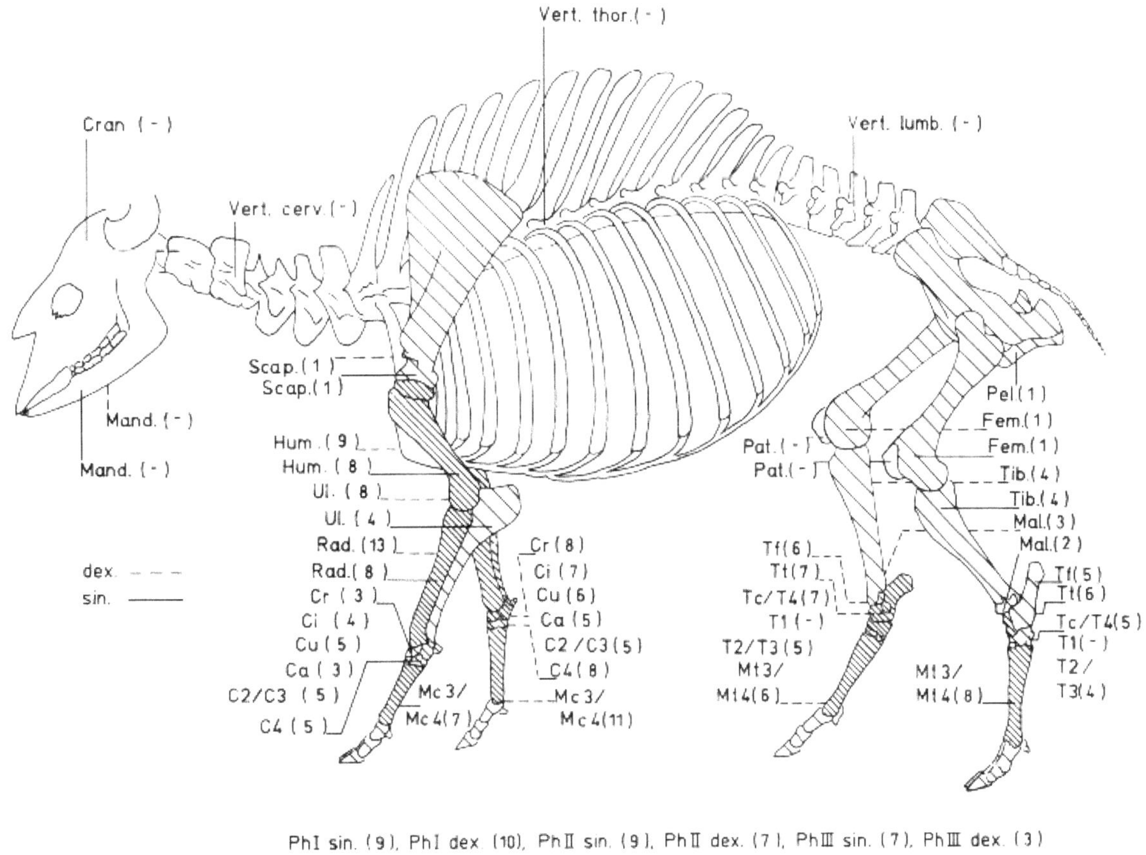

Fig. 60 *Bison menneri*, Untermassfeld.
Categorization of reconstructed anatomically connected units. In brackets: number of respective specimens.

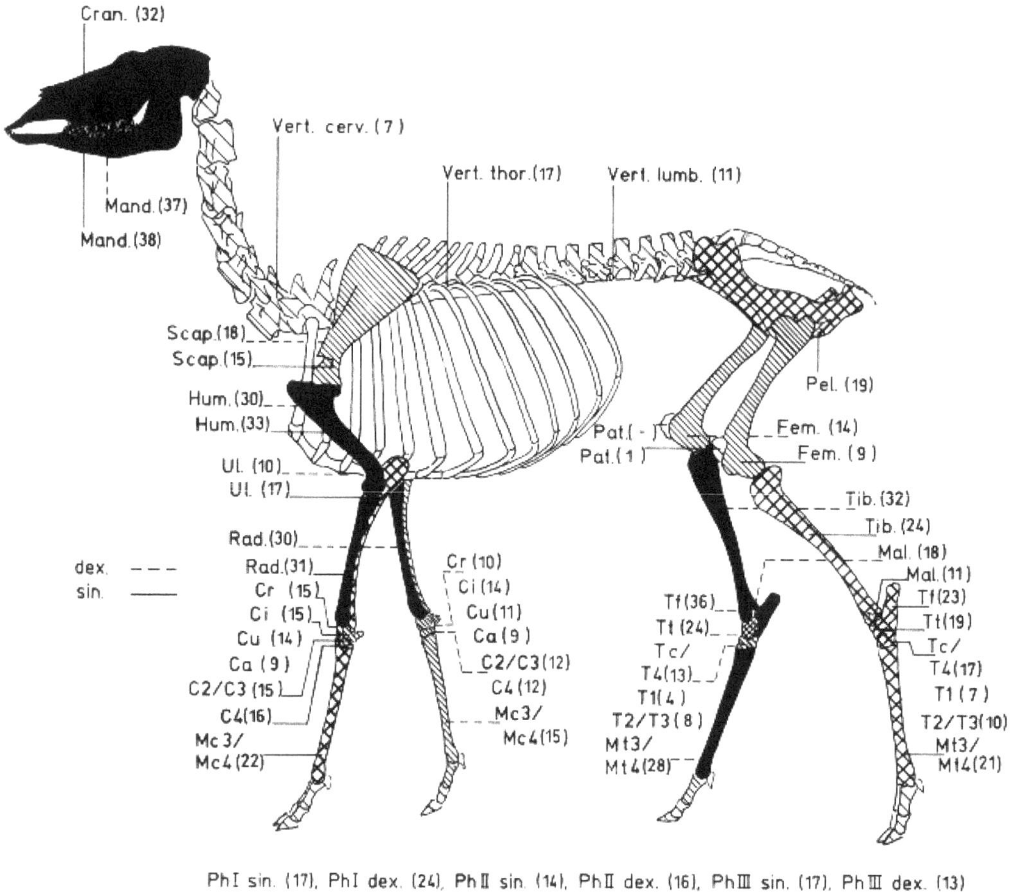

Fig. 61 *Eucladoceros giulii*, Untermassfeld.
Categorization of all skeletal elements recovered. In brackets: number of respective specimens.

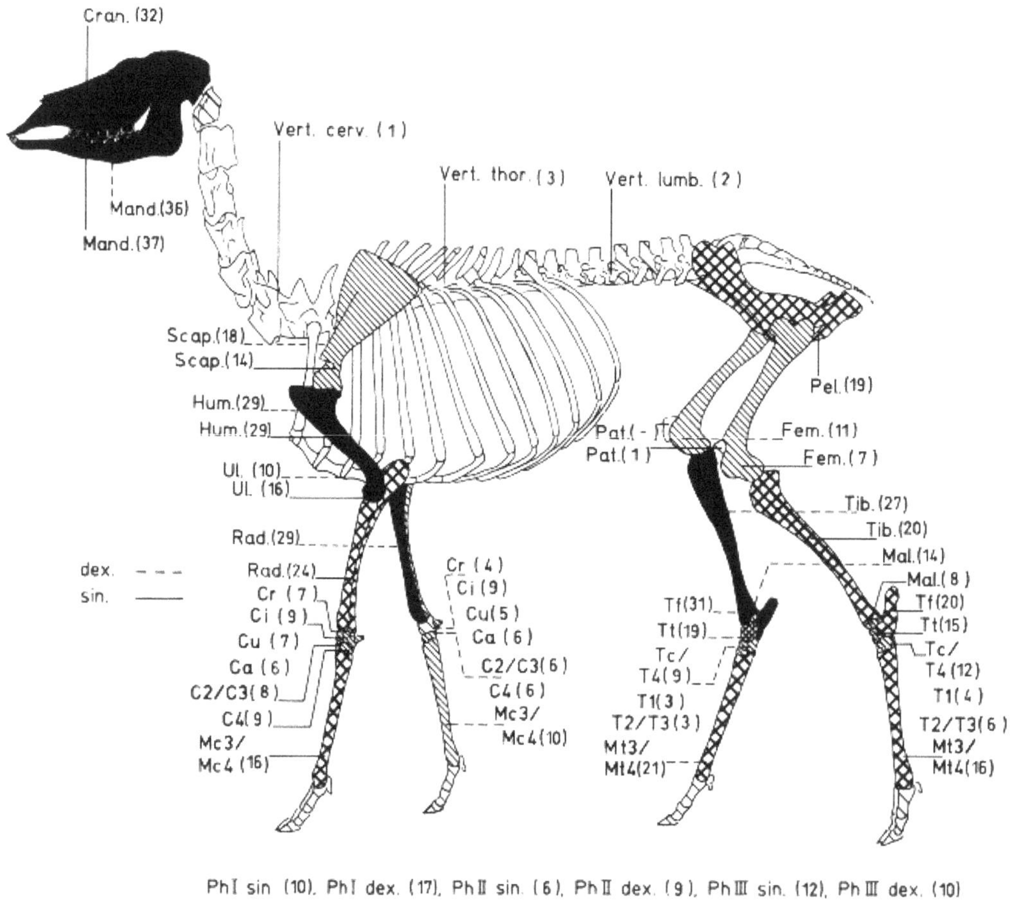

Fig. 62 *Eucladoceros giulii*, Untermassfeld.
Categorization of isolated skeletal elements. In brackets: number of respective specimens.

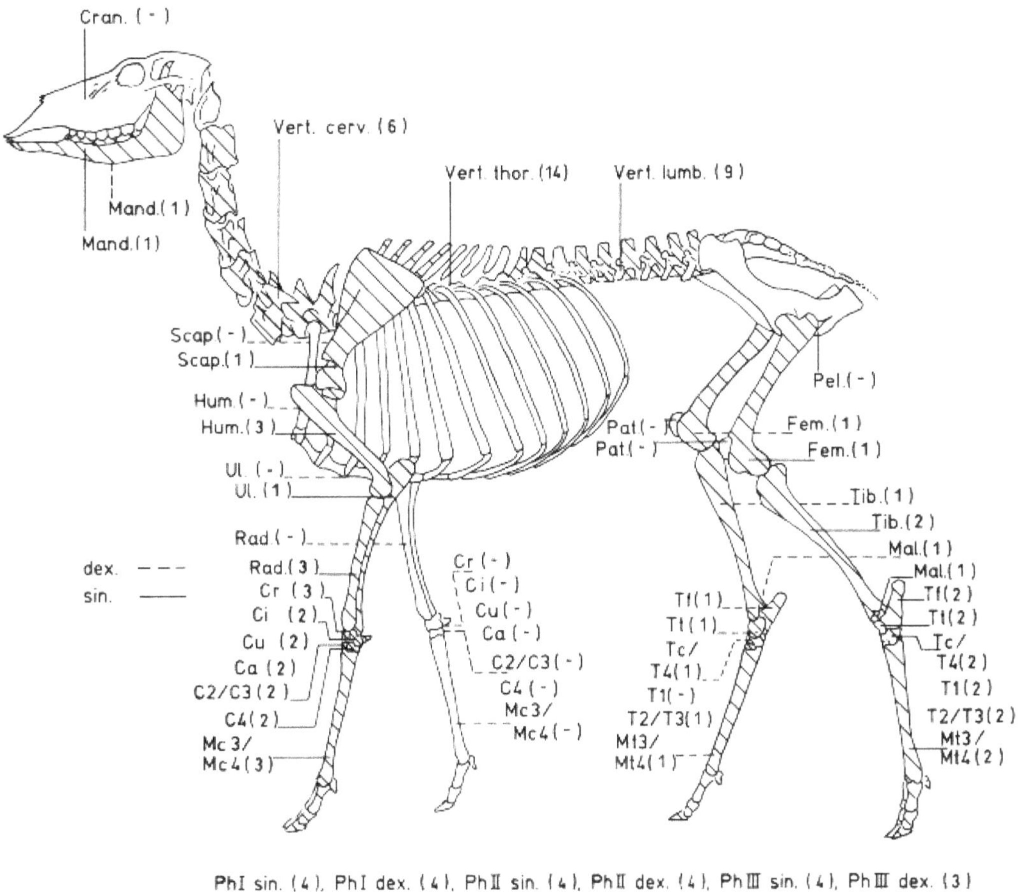

Fig. 63 *Eucladoceros giulii*, Untermassfeld. Categorization of anatomically connected units, and supposed anatomically connected units. In brackets: number of respective specimens.

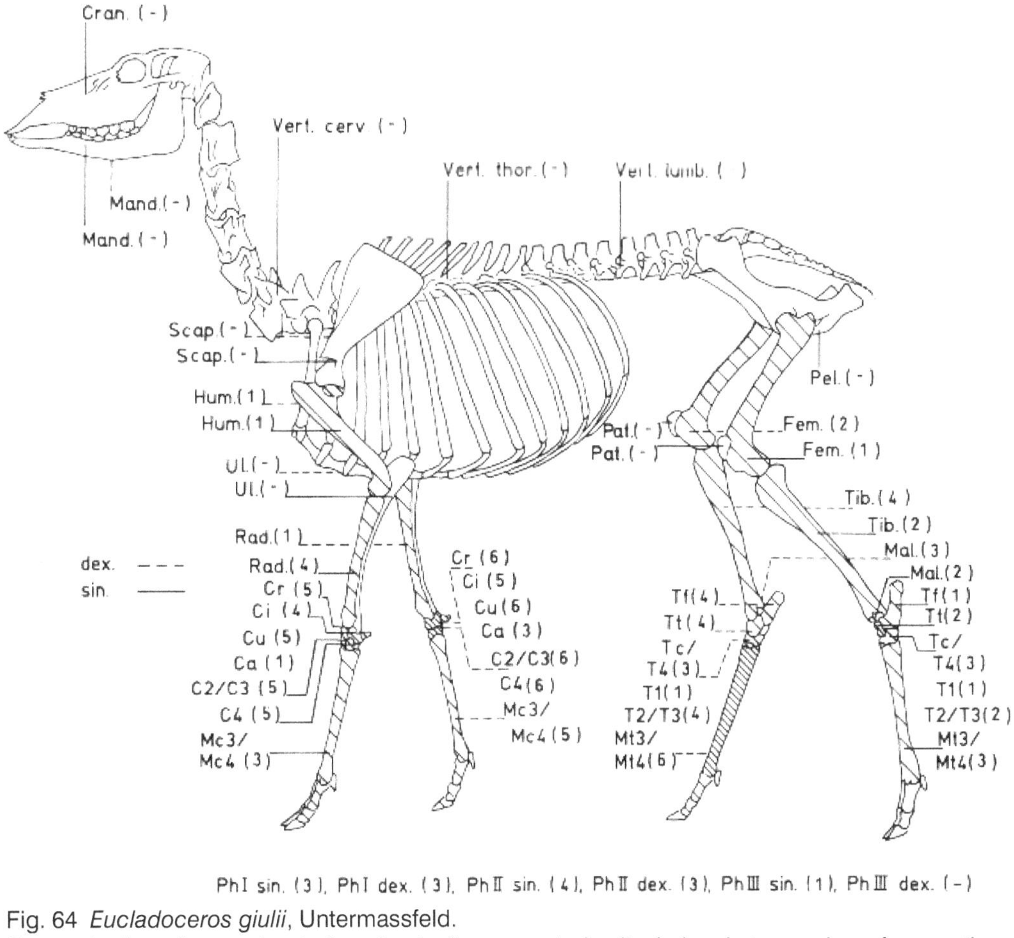

Fig. 64 *Eucladoceros giulii*, Untermassfeld.
Categorization of reconstructed anatomically connected units. In brackets: number of respective specimens.

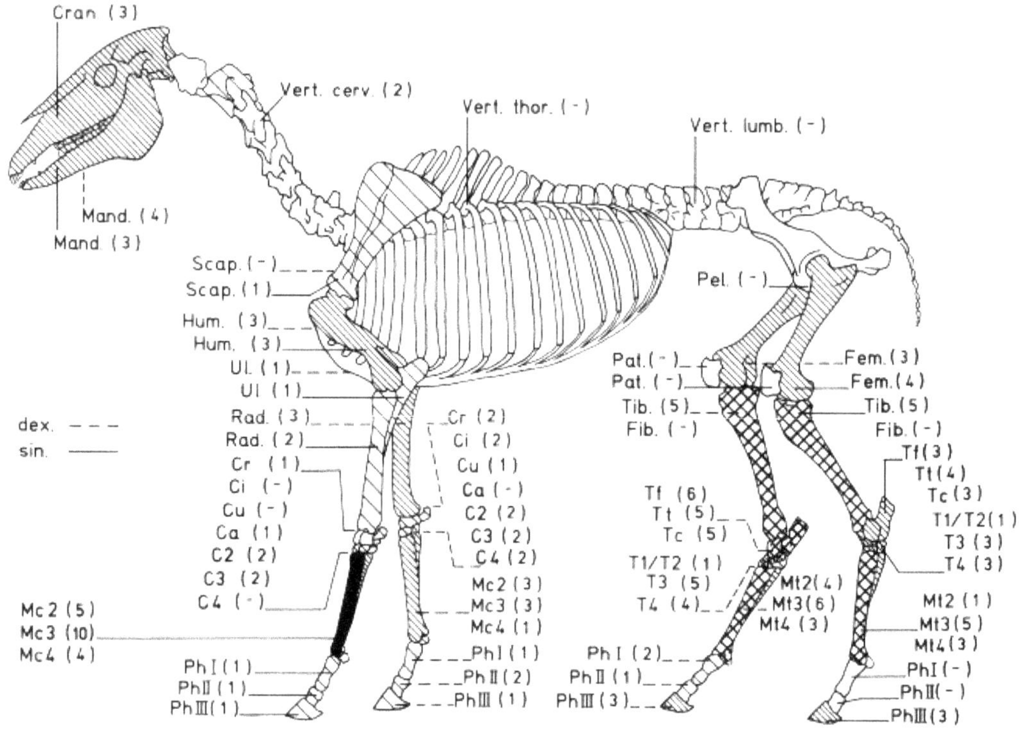

Fig. 65 *Equus wuesti*, Untermassfeld.
Categorization of all skeletal elements recovered. In brackets: number of respective specimens.

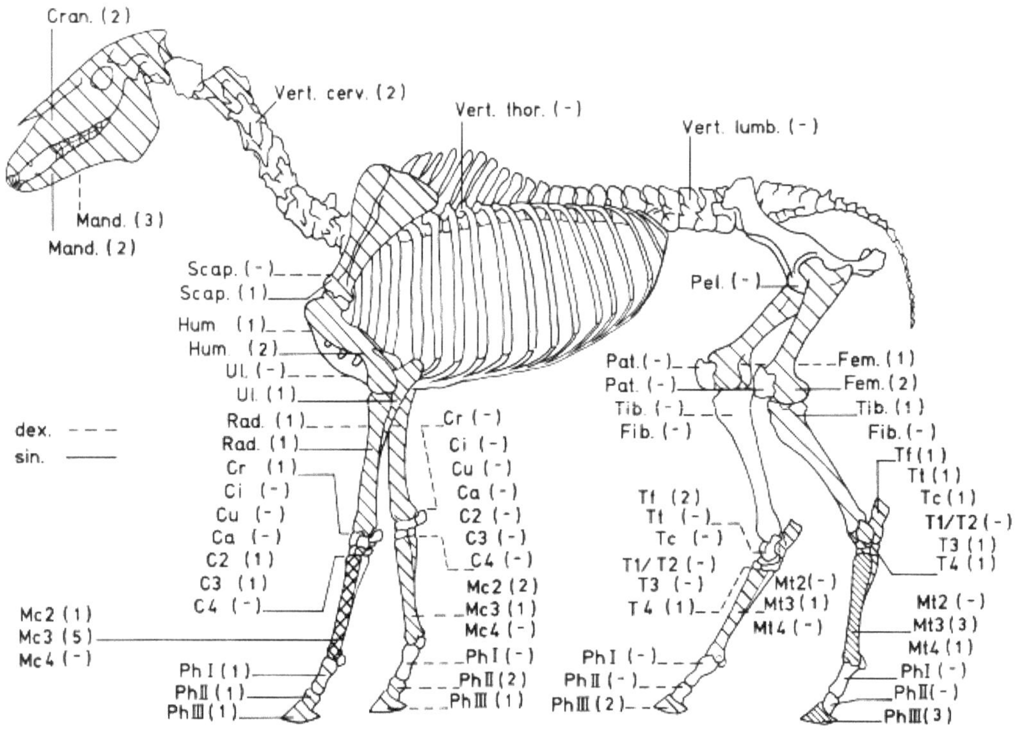

Fig. 66 *Equus wuesti*, Untermassfeld.
Categorization of isolated skeletal elements. In brackets: number of respective specimens.

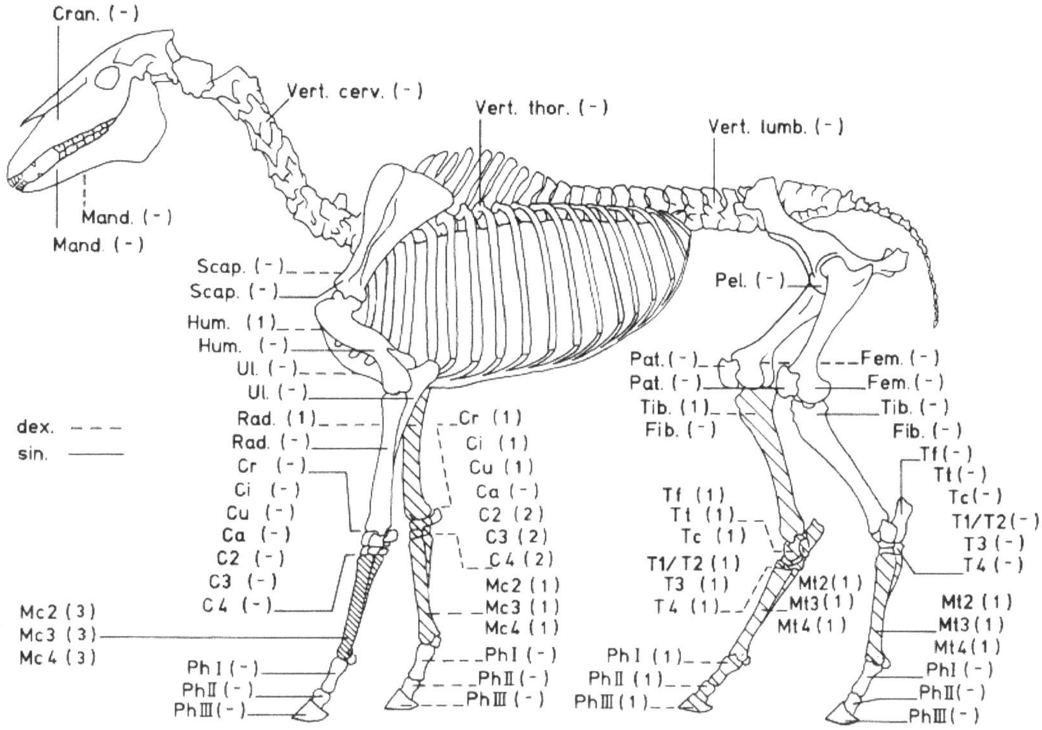

Fig. 67 *Equus wuesti*, Untermassfeld.
Categorization of anatomically connected units, and supposed anatomically connected units. In brackets: number of respective specimens.

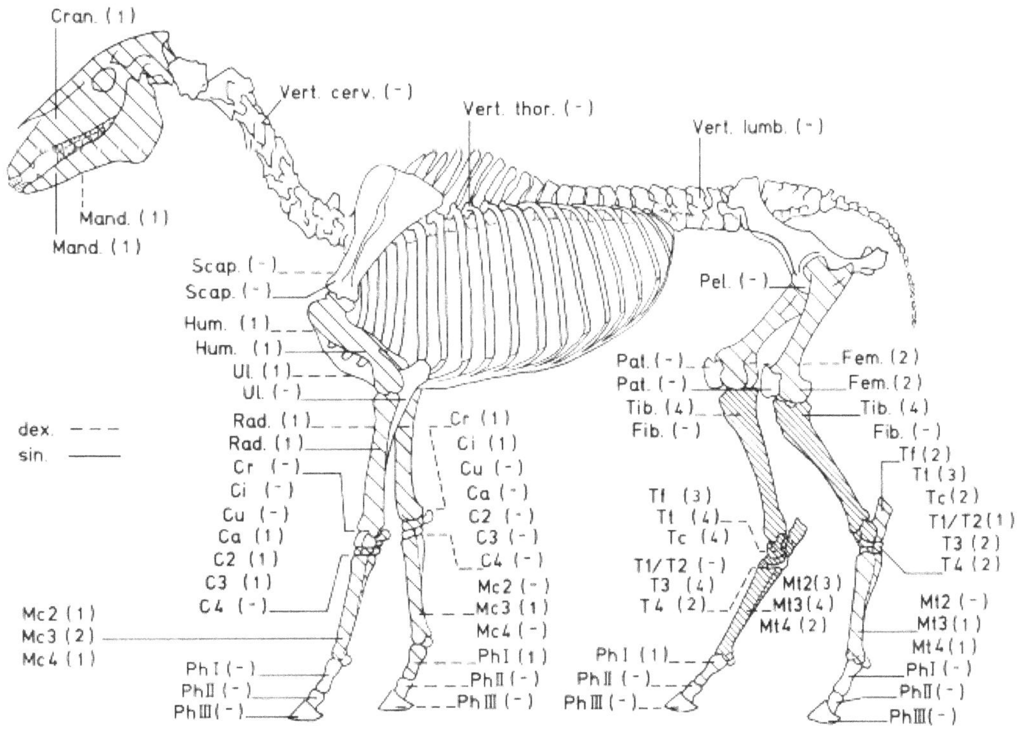

Fig. 68 *Equus wuesti*, Untermassfeld.
Categorization of reconstructed anatomically connected units. In brackets: number of respective specimens.

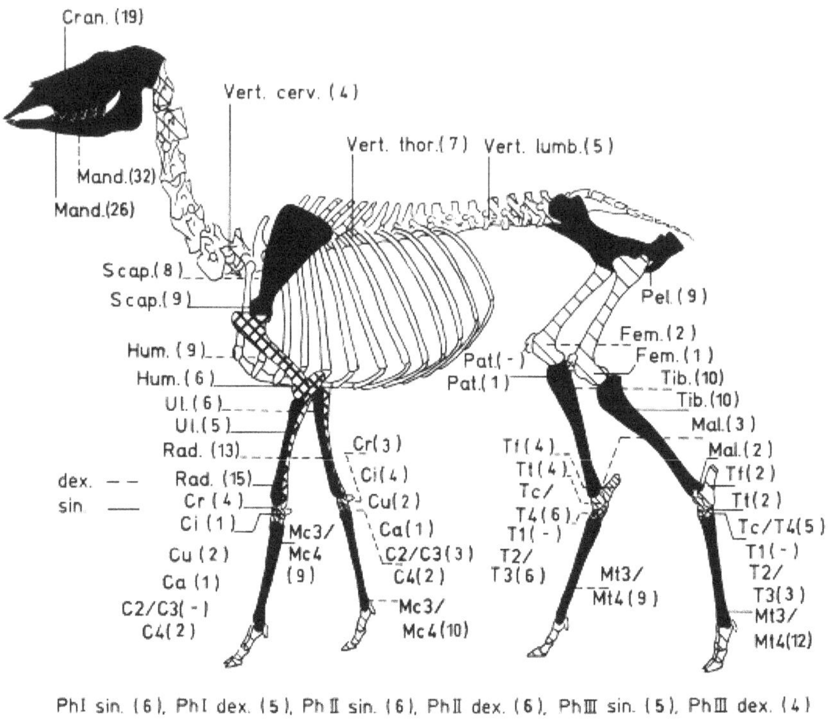

Fig. 69 *Cervus* s.l. *nestii vallonnetensis,* Untermassfeld.
Categorization of all skeletal elements recovered. In brackets: number of respective specimens.

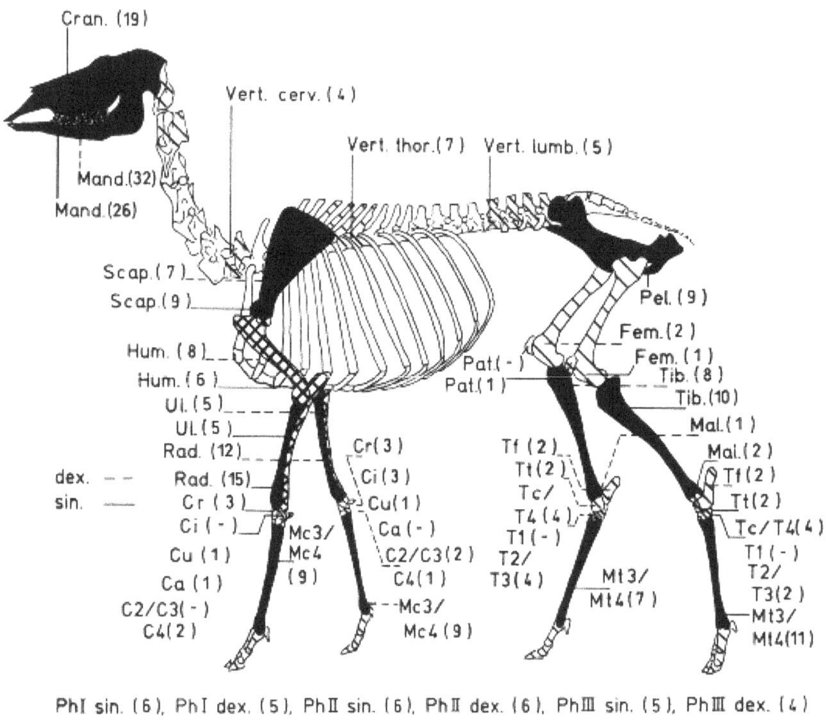

Fig. 70 *Cervus* s.l. *nestii vallonnetensis,* Untermassfeld.
Categorization of isolated skeletal elements. In brackets: number of respective specimens.

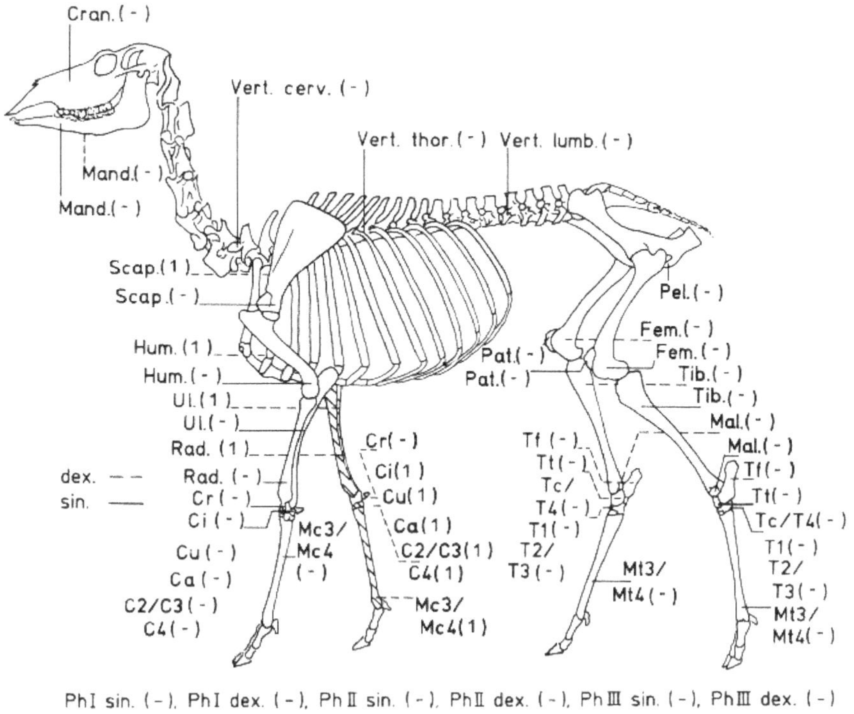

Fig. 71 *Cervus* s.l. *nestii vallonnetensis,* Untermassfeld.
Categorization of anatomically connected units, and supposed anatomically connected units. In brackets: number of respective specimens.

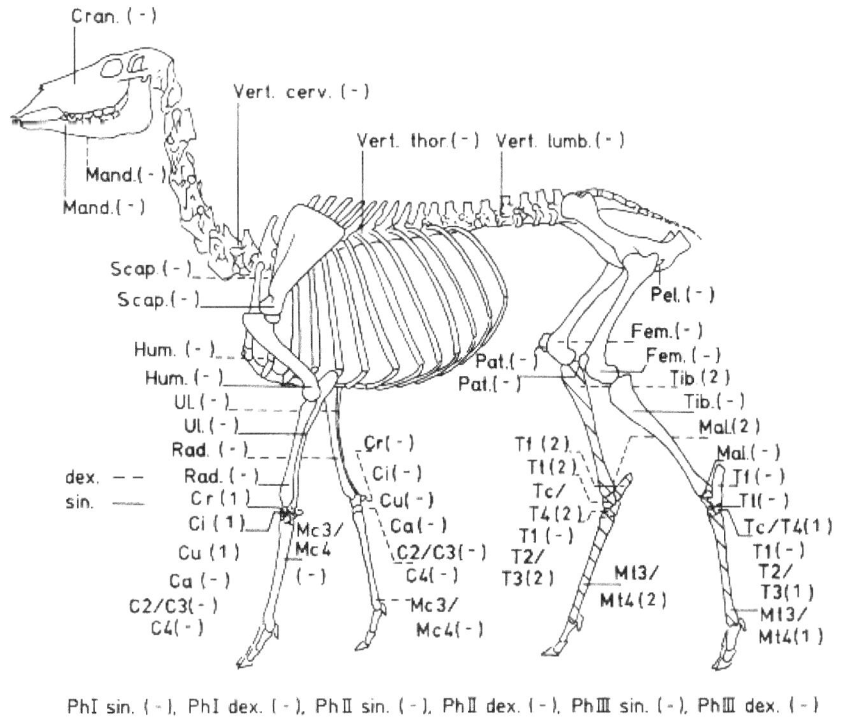

Fig. 72 *Cervus* s.l. *nestii vallonnetensis,* Untermassfeld.
Categorization of reconstructed anatomically connected units. In brackets: number of respective specimens.

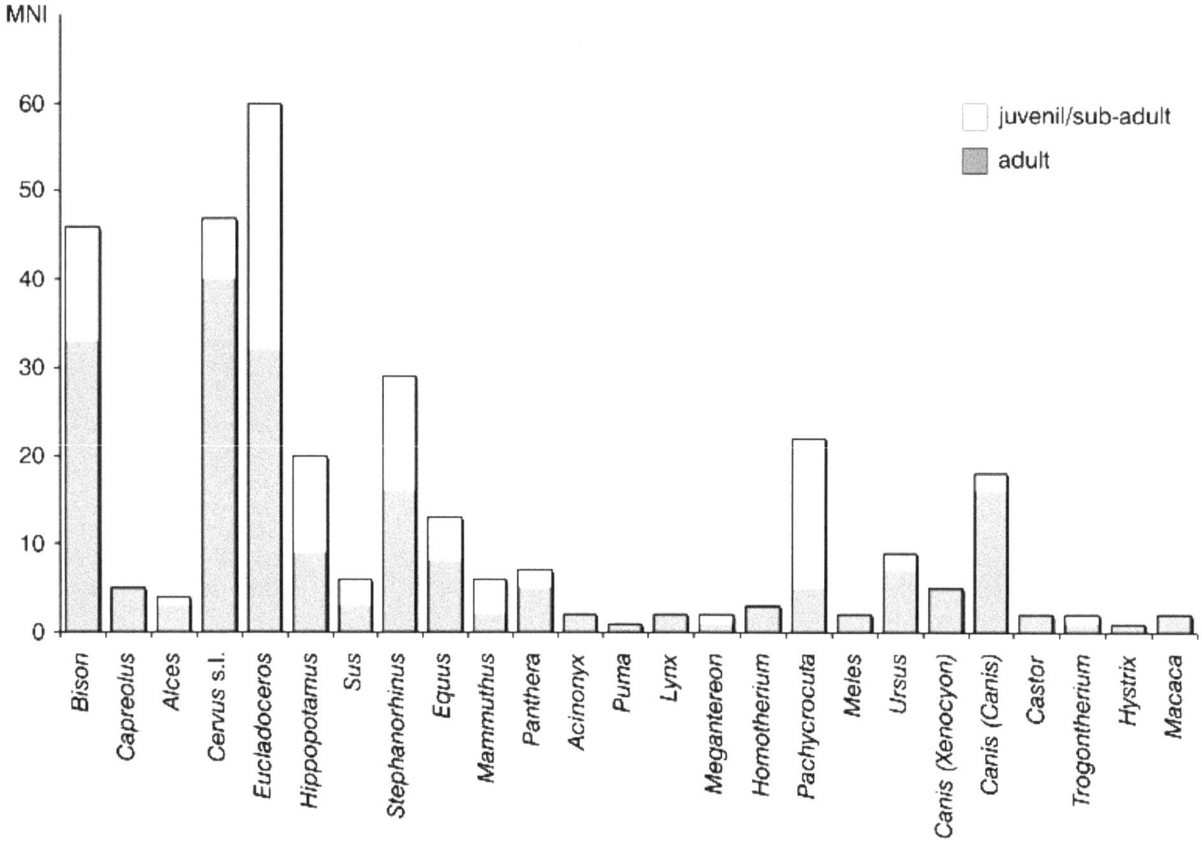

Fig. 73 Larger mammal community of Untermassfeld.
Distribution of individual age within the recovered larger mammal samples (based on MNI determined).

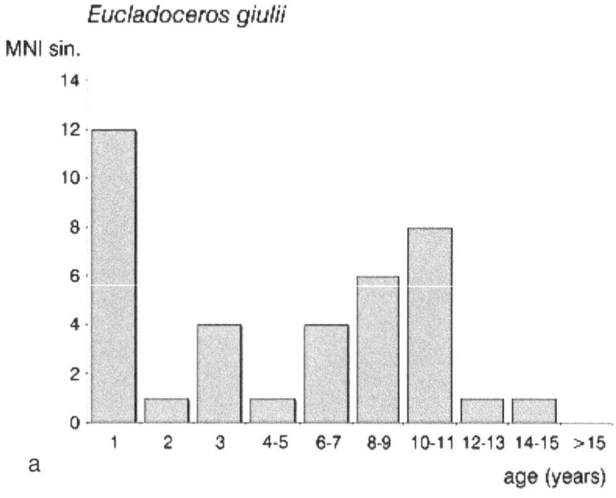

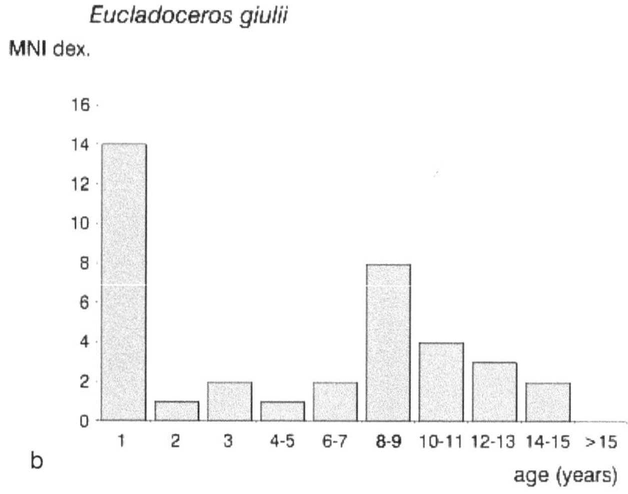

Fig. 74 *Eucladoceros giulii*, Untermassfeld.
Mortality pattern based on mandible fragments: a left. – b right.

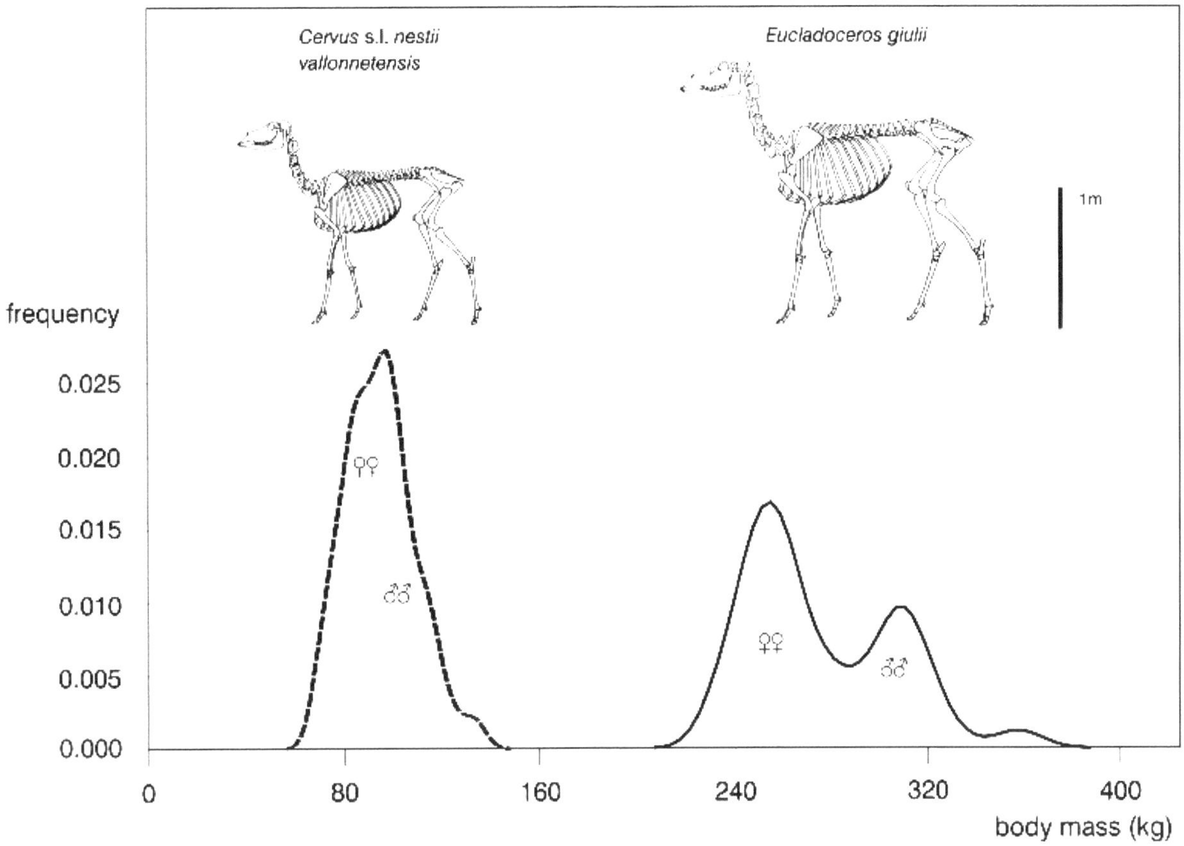

Fig. 75 *Cervus* s.l. *nestii vallonnetensis* and *Eucladoceros giulii* (adults), Untermassfeld.
Above: Sketch of body size relation based on complete long bones. – Below: Body mass range calculated from the lengths of the lower molar rows by using the allometry formula for cervids of Janis (2003).

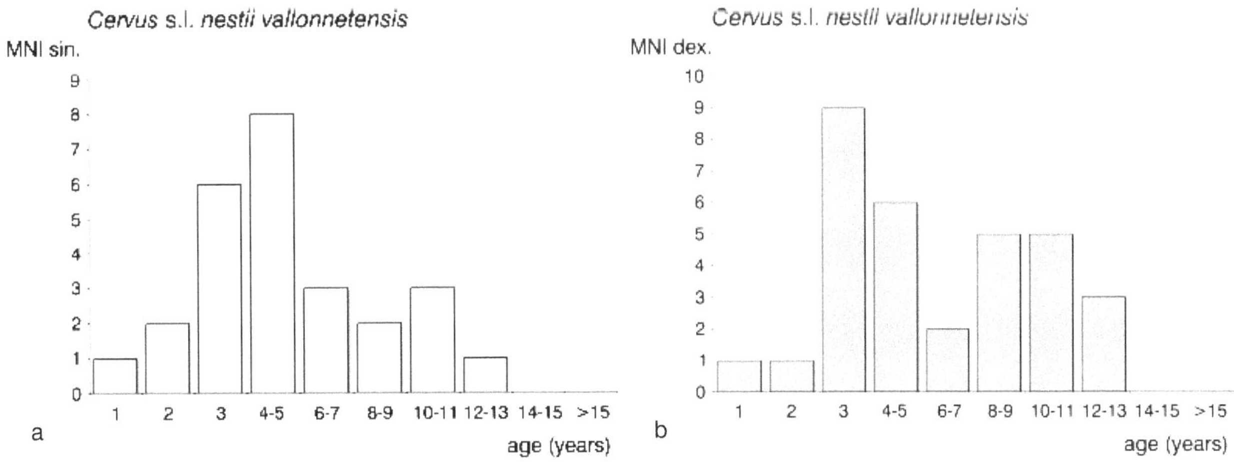

Fig. 76 *Cervus* s.l. *nestii vallonnetensis*, Untermassfeld.
Mortality pattern based on mandible fragments: a left. – b right.

5. Model of site genesis

5.1. Sedimentological and hydrological processes

For the fossiliferous parts of the Lower and Upper Fluviatile Sands of Untermassfeld, the following sedimentation processes can be deduced (Ellenberg & R.-D. Kahlke 1997; R.-D. Kahlke 2001d):

The accumulation of the skeletal remains from both the uppermost section of the Lower Fluviatile Sands (flooded surfaces) and the lower part of the Upper Fluviatile Sands (Channel Infill) took place in a spatially identical position under the lee of the Coarse Clastic Mudflow Fan that interfingers with the profile from the east.

With regard to the individual layers of the Lower Fluviatile Sands, lithological characteristics demonstrate repeated and largely similar, widely extended high flood events with conformable sedimentation rates (Fig. 77a). Several of these flood events carried carcasses or parts of these into the area of the present site. Today at least two fossiliferous accumulations, separated by a silt deposit of maximally 0.2 m thick (R.-D. Kahlke 2001d, 434ff., fig. 2), can be distinguished. One followed the other within a short period of time. Detailed taphonomical research – for example a computer-aided spatial representation of the find distribution combined with the reconstruction of individual skeletal units from the corresponding excavation levels – could reveal a further differentiation of the sedimentation process in the fossil bearing section of the Lower Fluviatile Sands.

Immediately following the deposition of the Lower Fluviatile Sands, renewed flooding with an uncommonly high erosive force scoured a considerable channel, which was instantly filled in as the transport energy decreased (Fig. 77b). As this was happening, and especially during the initial phase of the infilling, a large input of necromass, i.e. the totality of dead remains of flora and fauna transported by the river, occurred. A sudden increase in the flow rate of the run-off must account for the formation of the erosional channel, which was cut to a depth of 4 m. The increased flow rate was possibly caused by the coincidence in time of several meteorological-hydrographical factors, or by the influence of the Coarse Clastic Mudflow Fan itself, which in its effects could have been comparable to present-day hydrodynamic engineering for keeping waterways navigable. Here the construction of groyne systems also leads to a decrease in transport energy and conformable sedimentation rates near to the river bank, while the rate of flow in and around the more central channel is increased and the riverbed therefore excavated.

The geomorphological situation in the middle and upper valley of the Werra and the position of gravel terraces upstream allow a reconstruction of the potentially flooded area at the time the site came into being (Fig. 78). In all probability a total floodplain area of approximately 94 km^2 may be considered the source of the specimens concentrated in the lagerstätte. The catchment area was formed by the mountain chain of the Thuringian Forest SE of Schmalkalden, the Thuringian Slate Mountains, and parts of the Triassic terrain of SW Thuringia (heavy mineral analyses of Quaternary Werra gravels are to be found in Ellenberg 1975).

In large areas upstream of the site, the high flood events are likely to have caused erosion that effected a "clear-out" of the necromass present, which included vegetable material, isolated bones and remains of carcasses.

A special, and for the animals in the Werra valley catastrophic, influence on the development of the inundations was probably exerted by a constricted bend of the valley between Henfstädt and Leutersdorf, approximately 12 km upstream of the Untermassfeld site (Fig. 78). The "Needle's Eye" ("Nadelöhr)" at that spot, which is still effective today, is formed by massive and towering walls of Wellenkalk (Lower Muschelkalk). It is certain that this valley meander was already present during the Early Pleistocene, because the difference in height between the floodplain and the margin of the valley is almost 100 m here. From the recent widths of the valley at the geomorphological level of the Younger Weathered Coarse Gravels it follows that the Early Pleistocene flat-bottomed valley of the Werra narrowed from >1.5 km to < 1.0 km in the vicinity of the "Needle's Eye". This means that at the time the fossil lagerstätte at Untermassfeld was being formed, a narrow passage or "bottleneck" was already present and with increased water supply caused impoundment, and consequently an enhanced rise of the water level in the floodplain upstream.[10] The absence of Pleistocene terrace remains at the "Needle's Eye" (including its gentle southern slope) indicates that the gorge through which the water passed was exposed to strong lateral erosion during the Pleistocene. Increased flow rates and turbulence must be seen as its causes. Especially when the water level was higher than usual, these could also be fatal to the relatively good swimmers among the

[10] Even during the pre-Roman Iron Age the continuous danger of inundation in the said area influenced the layout of the road system, an example of which is to be found in Grasselt (1994, 172f.).

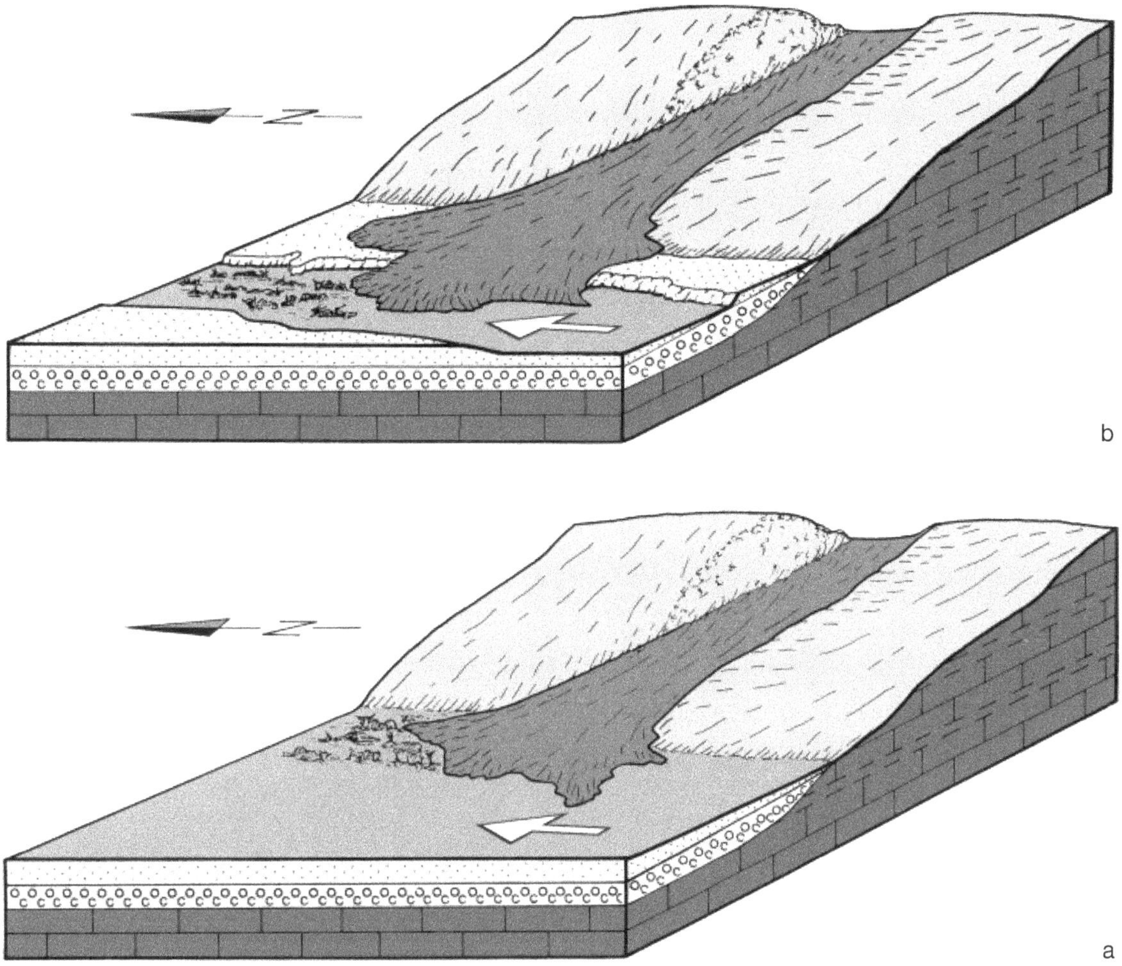

Fig. 77 Reconstruction of the area within which the Untermassfeld site was formed.
a At the time the Lower Fluviatile Sands were accumulated, accompanied by level deposition of high flood sediments. – b At the time the Upper Fluviatile Sands were accumulated and the erosional channel was filled in.

large mammals. During the high flood events that led to the formation of the concentrated lagerstätte of Untermassfeld, the wildlife in the valley of the Werra was therefore exposed to a dangerous situation, especially in the vicinity of the "Needle's Eye" between Henfstädt and Leutersdorf, as well as upstream of that area.

5.2. Origin and composition of the necromass

In view of the different ways of life and the biotope preferences of the animals present in the fossil lagerstätte (cf. Section 6.1.), the individual causes of death were probably very diverse. Against the background of a carnivore fauna that was rich in species and included many active hunters (cf. Section 3.2.6.; for the predator-prey ratios see Hemmer 2001) it is to be presumed that carcasses of prey in successive stages of decomposition were continuously present in the floodplain in the valley. Hemmer (l.c., 770) sees slightly less than half of the ungulate remains as related to carnivore kills upstream. Additional losses, for instance among juvenile cervids (see *Eucladoceros*), could have occurred during periods of increased wetness of the biotope, due to enhanced susceptibility to disease in a very wet understory. Moreover, by analogy with recent situations, a strongly species-related behaviour that caused the degree of endangerment to vary must certainly be taken into consideration. Exhausted and diseased animals, or those weakened by old age, were endangered in every respect.

For most species the relative abundance of right and left skeletal elements of the same individuals (cf. Section 4.6.) documents that largely intact carcasses were transported into the area of the site. Both this circumstance and the composition of the thanatocoenosis according to species and individuals indicate a not insignificant proportion of fresh victims of the inundation. Taking into consideration the proportion of carnivore kills

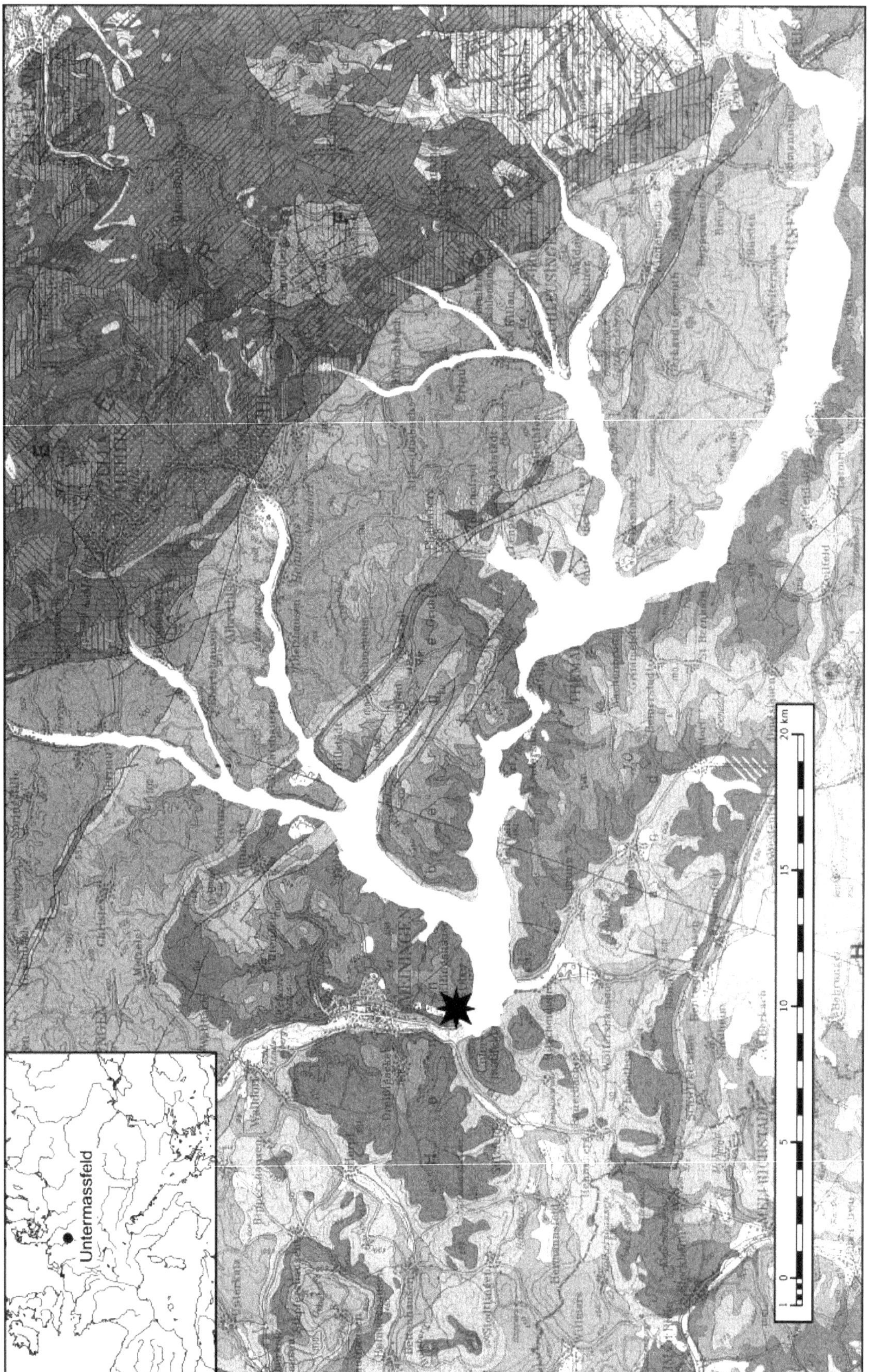

Fig. 78 Geography of the Untermassfeld site.
Potentially flooded area (blank) of the upper and middle regions of the River Werra and its tributaries during the genesis of the site. A narrow part of the valley some 12 km upstream of the fossil site (star) is clearly visible [Geological map after Seidel et. al. (1998)].

within the total body of material as estimated by Hemmer (see above), and assuming a certainly much smaller number of deaths by other factors, the figure that remains for the number of flood victims within the entire body of larger mammal material (MNI=313) is approximately 50%. For the vertebrate fauna the high floods in the Early Pleistocene Werra valley therefore turn out to have been a prominent cause of death, and, with regard to the fossil lagerstätte, also the decisive accumulative factor.

Of fundamental significance for the range, number of individuals, and mortality pattern of the vertebrate remains are the population densities of each species in the Early Pleistocene biotope, their specific behavioural patterns, and locomotory skills – especially swimming and climbing capabilities – and furthermore the way in which lairs were organized, and the weight or transportability of the carcasses that were generated (R.-D. Kahlke & Gaudzinski 2005). For comparison with the Early Pleistocene inundation scenarios for the Werra valley, artificially produced impoundment situations with rapidly rising water present themselves, especially within Africa, e.g., in the development of the Kariba reservoir in the Zambesi valley (Child 1968; Roth 1969, with further references). However, the at least temporarily rapid and energetic inundation of the floodplain along the Werra, as reconstructed from geological findings, probably had a more catastrophic impact on the fauna living there than the controlled construction of an artificial storage reservoir may have had.

In accordance with the geomorphological situations that exist in large inundated areas, island complexes of various dimensions are first formed, which the endangered wildlife then seeks out as life-saving refuges. If these are successively flooded, however, they must be abandoned. Depending on the number of individuals and the length of stay, constitutional weakening as a result of food shortage may already become observable in individual animals at this stage. Possible additional dangers that occur during flooding include poor visibility owing to rain or mist, the impossibility of passing along accustomed tracks, or panic reactions by stricken individuals.

In general, larger terrestrial vertebrates (including juveniles) at first actively keep their heads above the water when immersed in a flood, but differences in swimming capacities and stamina cause variation in the survival probability of the individuals or species. The apparently rather high number of recently killed flood victims within the entire body of material from Untermassfeld allows an interpretation of the individual age distribution from the remains of larger mammals (Table 9, Fig. 73). Partial similarities exist to the pattern that emerges from the threats posed by carnivores, whose kills also contributed a considerable quantity of bone material (see above).

The age and sex distribution determined for *Bison menneri* on the basis of the finds from Untermassfeld suggests a natural population and could indicate the perishing of gregarious groups of individuals. Comparable evidence of a modern mass drowning assemblage of bison (*Bison bison*) from Wood Buffalo National Park (Alberta, Canada) was reported by Haynes (1988, 226). In gregarious animals the individual readiness to cross deep waters by swimming is often subordinate to social bonds within the herd (cf., e.g., Roth 1969, 18).

Among the cervids, generally equipped with considerable swimming capacities, marked differences in the respective age class proportions are found for the more frequently recovered species (*Eucladoceros giulii*, *Cervus* s.l. *nestii vallonnetensis*) (elaborated in R.-D. Kahlke & Gaudzinski 2005). Very young and very old individuals of the large *Eucladoceros* are predominant, whereas the intermediate age classes are underrepresented. On the other hand, these are very well documented for the small deer, suggesting that a more feeble physical performance of *C.* s.l. *nestii vallonnetensis* became a threat to all age classes. In an analogous way, an entire herd of marsh deer (*Cervus duvauceli*) fell victim to floods caused by vehement monsoon rains along the Brahmaputra and Diphaloo in the Kaziranga National Park (NE India) in 1998 (Rai 1998, 22). During the inundation catastrophies, but also as targeted prey of the carnivore community, the strongest individuals of the Untermassfeld *Eucladoceros* in the intermediate age groups had a better chance of survival.

The mortality rate worked out for Untermassfeld juvenile *Hippopotamus amphibius antiquus* individuals on their milk dentition is clearly above that found for corresponding parts of recent populations (data in Eltringham 1999, 100-109). Modern observations show that a separation of calves from the female parent, brought about by the flow regime, is to be considered the most frequent cause of death. According to observations by Klingel (1995), hippopotamuses generally tend to escape high floods. The high proportion of young hippopotamus individuals within the Untermassfeld finds therefore probably indicates that they were exposed to danger in high flood situations. Should there be any flood victims among the adult *Hippopotamus* individuals from Untermassfeld, then they were most probably killed in the turbulence of the "Needle's Eye" (cf. Section 5.1.). Wild boar are only occasionally willing to cross wider expanses of water by swimming. However, the few suid remains so far available from Untermassfeld do not allow any more precise conclusions on population dynamics.

Stephanorhinus etruscus could not keep up swimming for long. Its capacities probably lay between those of the African black rhinoceros (*Diceros bicornis*) and the Asian rhinocerotids (*Dicerorhinus sumatrensis*, *Rhinoceros sondaicus*, *Rhinoceros unicornis*). All of these animals have trouble keeping their heavy heads above the water for long (Child 1968, 52). *Diceros bicornis* never goes into deep water voluntarily. When forced to do so, black rhinoceroses can swim short distances (10-25 m), but then go ashore again (Roth 1969, 15-17). Bathing Asian rhinoceroses are repeatedly observed, and sometimes they also cross fairly limited expanses of water, but while doing so may easily be drifted downstream (references in Child 1968, 52; Roth 1969, 17). In high flood situations they are much endangered. An example concerning *Rhinoceros unicornis* from the Kaziranga National Park (NE India) is given by Rai (1998, 22). The individual age profile of the Untermassfeld *Stephanorhinus etruscus* finds – as far as it can be determined – roughly corresponds to that of a well-researched, large recent population of *Diceros bicornis* from the Tsavo National Park in Kenya (Goddard 1970, table 8). In the smaller *D. bicornis* population from the Kariba basin (Zambesi valley) observed by Roth & Child (1968, table 1), the proportion of calves and juveniles turned out to be somewhat smaller than in the Untermassfeld material. In general it may be stated that juvenile to adult rhinoceroses of the Untermassfeld population were almost equally endangered by high floods.

Equids are likewise considered as slow, not very skilful swimmers. Unusually cold water combined with the effort of swimming may lead to a dangerous lability of the autonomic nervous system and end in circulatory collapse. *Equus wuesti* individuals of all age classes were to the same extent threatened by high flood situations.

Conversely, elephants have long been known for their powers of perseverance in swimming (see, e.g., the illustration in Chadwick 1991, 3-4). In various parts of the world, representatives of the proboscideans managed to reach islands at considerable distances from the mainland by actively swimming or drifting (examples in Bergh et al. 1996; Caloi et al. 1996). Especially the elephant's trunk, which can be used as a snorkel, and the trussed internal construction of the upper skull region (diploë) increase its chances of survival in water (Sondaar 1986). With regard to the speed of goal-oriented swimming, the performance of present-day African elephants (*Loxodonta africana*) is comparable to that of humans. Roth (1969, 15) describes how juvenile animals swim after their mother by clutching her tail. In accordance with their considerable swimming powers, only six individuals of *Mammuthus* have been documented from Untermassfeld. Three of these are juvenile to subadult animals. It seems possible that the elephants went down, not only through the flood water itself, but also because their inadequate climbing capacities did not enable them to scramble onto a life-saving bank.

In each case concerning the felid fauna – which was rich in species but naturally poor in individuals – it remains unclear whether the animal was seized by the mass of water pouring into the landscape of the floodplain or, alternatively, perished only after it had been attracted by flood victims and pounced on its prey. The affinity for water which is attributable to *Panthera onca gombaszoegensis* is similar to that of the recent jaguar (Hemmer et al. 2001), which also hunts in stagnant or slowly flowing waters (A. Turner & Antón 1997, 65). The lynx (*Lynx issiodorensis* ssp. ex aff. *spelaeus*) is likely to have put up much less resistance against the water. Because its breeding lairs are to be expected especially on the south-facing slopes (cf. also Hemmer 1993b, 1158) of the Werra valley, the absence of remains of juvenile animals from the thanatocoenosis is not surprising. The other cats whose presence was demonstrated (*Acinonyx pardinensis pleistocaenicus*, *Puma pardoides*, *Megantereon cultridens adroveri*, *Homotherium crenatidens*) probably also possessed a certain ability to swim, but turbulent floods overtaxed their powers.

Among the rather frequently demonstrated remains of hyaenas (*Pachycrocuta brevirostris*), juvenile animals with life spans shorter than one year clearly predominate (77.3%). Although corresponding age class proportions or mortality rates are hard to quantify for recent hyaena populations, observations of spotted hyaenas (*Crocuta crocuta*) in the Ngorongoro crater and the Serengeti national park (Tanzania) by Kruuk (1972, 35-36) and Jaarsfeld (1993) showed that under normal conditions mortality during the first year of life is lower than at Untermassfeld. This is also true of the Early Pleistocene assemblage of Venta Micena (Orce, Spain), which was formed by the transport and accumulation by *Pachycrocuta brevirostris* of carcasses and body parts of larger mammals close to breeding lairs dug by these carnivores (Palmqvist et al. 1996; Arribas & Palmqvist 1998; Palmqvist & Arribas 2001). Merely 40% of the *Pachycrocuta* individuals recorded from Venta Micena represent juveniles (Palmqvist & Arribas 2001, table 1). Furthermore, the vertebrate assemblages of the so-called hyaena caves of the European Middle and Late Pleistocene as a rule contain less than 50% juvenile individuals of the total MNI of *Crocuta*, e.g.: Lunel-Viel (Hérault, France) – 42.6%; Tornewton Cave (Devon, England) – 35.0%; Teufelslucken (Eggenburg, Austria) – 32.8%; Fouvent (Haute-Saône, France) – 26.5%; La grotte de la Carrière, Gerde (Hautes-Pyrénées, France) – 16.7% (data from Fosse 1995, 1996). With 68.1% the percentage of juvenile hyaenas in the Spanish Cueva de las Hienas (Asturias) exceeds these values (Fosse 1995, table 4), but is still lower than the percentage of juveniles recorded from Untermassfeld. It is therefore certain that hyaena dens with very young and hence much-endangered animals were present in the flooded area upriver from the present site of Untermassfeld. Perhaps the sandy river deposits of the valley floor were in some special way suited for making such structures. It cannot be ascertained whether there are also victims of intraspecific quarrelling within litters among the juvenile *Pachycrocuta* remains, but at least the possibility cannot be excluded. It is likely that adult animals were very good swimmers because of their physical strength.

In stark contrast to the hyaenas, remains of adult individuals of both sexes of *Canis mosbachensis* are predominant (88.9% of the total number of individuals), but they do not in any way reflect the age structure of a natural population either (cf. Smirnov & Korytin 1985; Peters 1993, 82). In this case the breeding caves or lairs as well as so-called "rendezvous places" of the young animals apparently lay beyond the inundated areas of the floodplain, possibly on the slopes of the Werra valley or on the adjacent plateaus (cf. Section 6.2). Because *Canis* was certainly not a bad swimmer, the high minimum number of individuals (18) inferred from the finds could indicate a fairly high population density of animals in the vicinity of the site.

The bear *Ursus* cf. *dolinensis* is also mainly represented by fully-grown animals, whereas remains of only two relatively young individuals are documented. Once in a while the animals certainly sought out shallow waters to catch fish, but it seems likely that high floods over larger areas often overtaxed their powers. Because the augmentation of the Untermassfeld thanatocoenosis largely took place during the winter half-year (cf. Section 5.5.), some of the identified bear individuals might have been surprised by water masses pouring in while they were hibernating.

Recent badgers (*Meles meles*) show a tendency to burrow their sets in south- to west-facing slopes (literature in Lüps & Wandeler 1993, 876-877), which could offer an explanation for the rarity of badger remains (*M. hollitzeri*) in the Untermassfeld fauna.

Since they are known to be excellent swimmers, it is not surprising that the occurrence of beavers (*Castor fiber*, *Trogontherium cuvieri*) is likewise extremely rare among the finds from Untermassfeld. They were hardly threatened by the high floods. Their remains are found more frequently in fine clastic deposits of silting-up river arms or lakes, which represent protracted sedimentation processes, as exemplified by, e.g., the Tegelen Clays (Schreuder 1929; Mayhew 1978). The macaques (*Macaca sylvanus*) were able to escape from the water masses effortlessly, thanks to their climbing capacities.

The listing of those mammalian taxa that have so far not been identified in the Untermassfeld fossil community but may certainly be expected on the basis of the stratigraphical and ecological situation supports the idea that high flood events were of causal significance for the genesis of the site. Thus it is conceivable that good climbers, such as the Early Pleistocene European wild cat (*Felis lunensis* or *F. silvestris lunensis*), the ermine / stoat (*Mustela palerminea*), the weasel (*Mustela praenivalis*), and probably also the beech marten (*Martes vetus*), which climbs well, were often able to rescue themselves onto branches or twigs above the water level[11]. It is also conceivable that natural rafts allowed lightweight mammals to survive. If *Pannonictis*, that may have swum, or the common otter *Lutra,* lived in the Early Pleistocene Werra valley, they, too, were hardly threatened by high flood events.

It seems likely that representatives of species that live in burrows – such as some mustelids (e.g., *Mustela palerminea*, in part also *M. praenivalis* and *Martes vetus*) and foxes (*Vulpes alopecoides* or *V. praeglacialis*[12]) that have so far not been found among the recovered material either – were drowned and embedded in these very burrows by water masses rushing in suddenly.

The Untermassfeld thanatocoenosis mainly lacks small species with body masses below 10 kg. Young animals that belong to larger species and are but marginally heavier are likewise underrepresented. Both as prey of carnivores and in case of fluviatile transport, carcasses in this weight category are not only destroyed much faster than bigger vertebrate bodies, but also to a higher degree (cf. Section 4.6, Table 8). In addition, they are more easily filtered out of running water by shrubs than bigger bodies or their still sizeable parts.

[11] For the Central-European stratigraphical range of the taxa mentioned, see Wolsan (1993).

[12] The validity of the taxa mentioned is discussed in R.-D. Kahlke (1994, 14-15).

Remains of small vertebrates from the site are again present in fairly large numbers. For concentrations of small vertebrates recovered during a principal digging period (1983-1989), Maul (2001a, 859) stated an average of 23 finds per ton. The remains of dentitions and skeletons originate from disintegrated pellets or droppings of the respective hunters as well as from animals that perished in expanses of open water. The number, species range and osteological diversity of the finds result from various ecological and taphonomical selection processes (Maul 2001a, 858-861). It seems possible that the small mammal skeletons (mainly *Spermophilus*, cf. Section 3.2.6.) discovered in close vicinity to the Coarse Clastic Mudflow Fan that wedges into the fossiliferous deposit were derived from destroyed animal burrows within the slumped sediment body, especially because the presence of fully open landscape components at the time of the formation of the fossil assemblage can be demonstrated for the adjacent south- to southwest-facing slope surfaces (cf. Section 6.1.).

Only the smallest part of the rather incomplete evidence of the Untermassfeld fish and bird faunas is probably to be traced to flood victims. Skeletal elements that belong to a single raven (*Corvus* sp.) described by Jánossy (1997, 98-99) suggest an individual embedded immediately after its death. Because of the effects of their feathers, air cavities, and large relative surfaces, bird carcasses in general may stay afloat for a fairly long time, up to 40 days according to Schäfer (1976). In this characteristic they differ from mammalian ones, so that the probability of a joint deposition and embedding in a fluviatile environment is reduced.

5.3. Accumulation and distribution of the necromass

The carcasses, or remains of them, that were carried into the area of the present site from the upstream section of the inundated Werra valley, gradually decomposed during transport. Apart from hydrodynamic pushing or pulling forces of the running water, collisions with natural obstacles and driftage also contributed to this process.

During the formation of the Lower as well as the Upper Fluviatile Sands, necromass was accumulated under the lee of the Coarse Clastic Mudflow Fan directly adjacent to the excavation area (cf. Section 2. and Figs. 77a, b). Differences in taphonomical features specify the characteristics of the respective sedimentation processes as deduced from the geological findings:

Compared to those within the Channel Infill (Upper Fluviatile Sands), the find concentrations within the fossiliferous layers of the Lower Fluviatile Sands are mostly smaller. The degree of disarticulation is lesser there, and skeletal elements are but little fragmented, which indicates a largely tranquil deposition of sediments in calmly flowing but high-level flood water, including the necromass carried along. Erosional gullies of little extent are to be looked upon as the results of runnels that cut into the sand surfaces present at the time. A bone concentration with an extremely high find density, clear orientation of long axes and a high proportion of articulated bone associations was discovered at the eastern border of the excavated area. It could indicate a washout seam that formed in shallow water during a longer period.

The distribution of finds within the Channel Infill (Upper Fluviatile Sands) of Untermassfeld reflects a process of variably intensive necromass influx. As the sediment began to be filled in, deposition of extremely high quantities of largely disarticulated, and in part fragmented, skeletal elements took place. This documents the high concentration of (partial) carcasses in the run-off immediately after the climax of a flood catastrophe. Skeletal elements that belonged to the same individuals, e.g., those of a hippopotamus (R.-D. Kahlke 2001e), were disarticulated and drifted apart, which illustrates the power that the running water in the channel still had.

As the infilling of the erosional channel continued, the number of vertebrate remains embedded was reduced. A more tranquil sedimentation now caused a certain regularity in the distribution of finds (cf. Section 4.3.1.). In keeping with the conditions set by the current, the main mass of heavy carcass portions was deposited in the central part of the area where the finds were made, about 10-22 m north of the Coarse Clastic Mudflow Fan. Directly north of the mudflow fan, however, the influx of small, light-weight, and in part strongly fragmented bones was predominant. Apart from the regime of the current, undulating wave movements close to coarse clasts apparently played a role in the fragmentation of the fossil material within this range. Nearer to the northern and northeastern border of the investigated area (cf. Figs. 41, 42) the accumulating effect of the Coarse Clastic Mudflow Fan weakened. The degree of articulation as well as the average size and weight of the separate finds gradually decrease.

Remains of carcasses were only occasionally embedded in the upper layers of the Channel Infill. An overlying, up to 0.2 m thick, fine-grained horizon without bones (the Silt Band) marks the end of sedimentation. Its high pelite content was deposited from particles suspended in the now largely calmed-down waters of the river ("dregs").

As already described, the processes of decomposition and relocation of skeletal remains within the area of the site were determined by running water and in principle took place in the order: Partial Skeleton (articulated find) – Separate Element (isolated find) – Accumulation of Finds (polyspecific bone concentration) (cf. Section 4.4. as well as R.-D. Kahlke 1997c, 393ff., 1999). With regard to the highly fossiliferous layers, the primarily excellent state of preservation of nearly all vertebrate remains from the site (cf. Section 4.2.) indicates high sedimentation rates, as expected when flood deposits are involved.

5.4. Modification and decomposition of the necromass

The decomposition of the skeletal material at Untermassfeld was furthered and modified by activities of the multifarious carnivore community of the Early Pleistocene biotope (cf. Section 3.2.6., 4.5.). These took place both before the carcass remains were carried into the area of the present site and at this very location. The estimated relatively high proportion of carnivore kills in relation to the total body of material (cf. section 5.2.) is of course reflected by the remains washed in, but intensive chewing by carnivores also occurred in those parts of the site that showed an increased number of carcass portions. This is illustrated by the reconstruction of a rhinoceros individual from isolated finds that, in accordance with this view, bear tooth marks (H.-D. Kahlke 2001b, 541, 544f.). Examples of dismemberment that occurred on the spot can also be given for other species. The regularly placed tooth marks were caused mainly by the strong *Pachycrocuta brevirostris*. Tooth marks by canids probably also occur, but are of subsidiary importance (cf. Section 4.5.). Occasionally there was an intensive preoccupation with isolated bones, which is reminiscent of activities by juvenile animals, possibly beyond the area of the site (R.-D. Kahlke 2001f, 900ff.). Detailed research may serve to demonstrate other causal agents.

In the area of the present site, the carcass portions were obviously temporarily within easy reach of representatives of the carnivorous species mentioned, in spite of high sedimentation rates. This is indicative of expanses of shallow water, possibly also of surfaces that fell dry briefly and partially. Another indication of sediment surfaces that occasionally dried up is the excellent state of preservation of the recovered coprolites of *Pachycrocuta brevirostris* with their surface morphology intact, as well as the occurrence of entire coprolite nests. Because it is unlikely that droppings could be transported by flowing water without incurring damage, the majority of the specimens, especially the nests that were found, should be interpreted as remains of fossil latrines that presumably served to mark territories (Keiler 2001c).

Some gnaw marks observed on the bone material were produced by small mammals (especially by *Hystrix* sp.), and the fact that they were made likewise presupposes the absence of water cover (Maul 2001b). Both the excellent state of preservation of the gnawed specimens, and their concentration within just a few plan squares of the excavation area, speak against their having been washed in from inundated areas upstream of the site. After the manuscript of Maul (2001b) had been finished, another eight gnawed bones were found very close to one another in sands of the Channel Infill (Q 785). On the other hand, transposition of a specimen from the nearby floodplain landscape has to be assumed for isolated evidence of osteophagia by herbivores (R.-D. Kahlke 2001f). The relative rarity (for example with respect to Pleistocene cave sites) of Untermassfeld material gnawed by small mammals could indicate that sediment surfaces that fell dry completely – a prerequisite for the penetration of small mammals into the area of the site – were only available for short time spans.

Until now, the Untermassfeld material has not been searched for possible marks of insolation, for instance for indications of intensive defatting and the ensuing superficial embrittlement of bone parts that were very radically exposed to the radiation of the sun.

Sintered root tubes now and then found in natural position[13] (Ellenberg & R.-D. Kahlke 1997, 45, pl. 14, 2), etched traces of a fine root network on bone surfaces (R.-D. Kahlke 1997b, 282, pl. 61, 9), as well as etched tubes caused by roots within coprolites (Keiler 2001c, 695f., pl. 131, 3-4) demonstrate the existence of vegetation cover – presumably for the most part herbaceous – at times when there was no sedimentation going on. Under not too dry conditions, the well-manured surface of a field strewn with carcasses is very rapidly overgrown with plants (cf. also Weigelt 1927, 21). At present, however, the time-lag between the deposition of the fossils, and the development of the vegetational community or communities, is not exactly determinable.

[13] In a wider context especially in Lower Fluviatile Sands within the area of the section called „Schwalbenwand" („Swallow Wall") (see R.-D. Kahlke 2001d, supplement I).

5.5. Seasonality

At the moment, statements regarding the season at which the site came into being may most clearly be inferred from the antler finds (H.-D. Kahlke 2001a, 474f.). 89,7% of the antler remains of *Cervus* s.l. *nestii vallonnetensis* are attached to skulls (pedicles without beams included). As far as can be determined, the recovered antlers of *Eucladoceros* and *Capreolus* are exclusively attached to skulls. Hence the period during which the carcasses were embedded may be narrowed down to time spans between October and the end of March or the beginning of April, i.e., always to winter half-years.

In order to determine at what age and in which season the animals died, H. Kierdorf (Cologne) performed root cement analyses on isolated cervid teeth. They failed, however, because diagenetic recrystallizations occurred in the samples studied. New trials on bear teeth are being prepared by N. Garcia (Berkeley / Madrid).

5.6. Postgenetic processes

The deposits immediately overlying the "Silt Band", i.e., the upper section of the Upper Fluviatile Sands of Untermassfeld, are again composed of sandy high flood sediments of the Werra (Figs. 2, 3). Here the accumulating effects of the Coarse Clastic Mudflow Fan already no longer existed, and finds of fossils have not been made at this stratigraphic level. The groundwater-impounding effect of the fine-grained Silt Band, and a neotectonically induced southwestward dip of, especially, the valley-facing segment of the entire sequence of strata, triggered intensive solifluction processes within the whole upper section of the Upper Fluviatile Sands (Ellenberg & R.-D. Kahlke 1997, 43ff.).

As the Early Pleistocene level of the Werra valley subsided (Ellenberg & R.-D. Kahlke 1997, fig. 2; Ellenberg et al. 1997, fig. 1), a sequence of cover deposits composed of escarpment loam with coarse scree layers intercalated at the side of the slope developed on top of the Fluviatile Sands of Untermassfeld. This sequence reflects sedimentation processes that took place during the span between the latest Early Pleistocene and the Holocene.

All in all, the fossiliferous parts of the Fluviatile Sands of Untermassfeld reflect a multiple sedimentation sequence determined by high flood events. The composition and grouping of the recovered fossil assemblage were determined by manifold topographical, climatological, biological, and taphonomical circumstances and factors, effective both in the wider surroundings of the site – especially on the upper reaches of the Werra – and directly in the deposition area itself. In their stratigraphical and taphonomical context, the preserved vestiges of life document the development of a concentrated palaeontological lagerstätte that formed within a geologically short and hence biostratigraphically instantaneous time span.

6. Palaeoenvironment

Riverside landscapes with their attractive wet areas and expanses of open water are not only inhabited by animals limited to the floodplain but are also regularly visited by species from the wider surroundings, for watering and in part for foraging. It has to be assumed that the floodplain landscapes of the Early Pleistocene Werra valley were likewise frequented by faunal elements from biotopes at greater distances. The high flood events deduced from the geological and taphonomical record (cf. Sections 2., 4.) could therefore carry remains of the living world of a larger area into the site.

6.1. Attribution to biotopes

The fish remains found at Untermassfeld (*Esox lucius*, aff. *Gobio* sp., *Scardinius* sp.) easily fit into the picture of a fluviatile fossil lagerstätte, but in no way represent the entire range of the local ichthyofauna at the time. At any rate, the presence of eight fish species was demonstrated for the early Middle Pleistocene clays of Voigtstedt, for example (Deckert & Karrer 1965). The early Middle Pleistocene of Ürömhegy (Hungary) counted ten species[14] (Beringkey 1959), the Tegelen Clays yielded eleven (Gaudant 1979), and the early Middle Pleistocene of Little Oakley even twelve (Lister et al. 1990). The Late Pliocene / Quaternary fish faunas of Germany were listed by G. Böhme (1997b).

Because of the size of its bones, the pike (*Esox lucius*) is most frequently documented at Untermassfeld. As a stationary fish of shallow water, it lives in stagnant or slow-moving, not too turbid waters (G. Böhme 1996, 29). For egg deposition it prefers shallow waters abundant with plants, or flooded fields. A rudd-like species (*Scardinius* sp.), a fish that swims in schools, likewise occurs in waters which are still or characterized by moderate flow rates. It constitutes the staple diet of relatively young fishes of prey (G. Böhme 1996, 31).

For the flat-bottomed valley of the Early Pleistocene Werra one has to assume the presence of blind arms of the river showing various stages of silting up, oxbow lakes and pools of all sorts as well as humid areas with boggy soils. Remains of newts (*Triturus* cf. *cristatus*, *T.* cf. *vulgaris*), tree frog (*Hyla* cf. *arborea*) and moor frog (*Rana* cf. *arvalis*) demonstrate the former existence of such biotopes. The smooth newt (*T. vulgaris*) prefers open, sunny pools. On the other hand, during the summer months the great crested newt (*T. cristatus*), which is known as an able swimmer, is sooner found in larger and hence cooler waters. The tree frog (*Hyla arborea*), when it has left its winter quarters, mainly stays in the vicinity or near the surface of swamps, pools or ponds. Outside the mating season, the moor frog (*Rana arvalis*) lives a nocturnal life on boggy fields near water, in peat bogs and humid deciduous forests in the floodplain.

It is not surprising that skeletal remains of a swan [*Cygnus* sp. (cf. *olor*)], a large goose [*Anser* sp. (aff. *subanser*)], and a sea-eagle [*Haliaeëtus* sp. (aff. *brevipes*)] were recovered from the site, because open expanses of water certainly existed there. The mute swan (*Cygnus olor*) needs large waters with abundant vegetation for feeding (upending). Its large nest is built right next to the shore. Feeding by upending, albeit combined with longer stays in marshy lowlands, must likewise be assumed for the goose. The sea-eagle nests on cliffs or old trees and feeds on fish and water fowl, but it also takes carrion, which was profusely available in the Werra valley as this was both full of carnivores and constantly threatened by high floods.

Further faunistic evidence of stagnant or moderately flowing waters is presented by *Mimomys savini*, *Sorex (Drepanosorex)* ex gr. *margaritodon – savini* and *Beremendia* cf. *fissidens,* as well as by the beavers *Trogontherium cuvieri* and *Castor fiber*. These beaver species had similar body sizes and were well adapted to a semiaquatic life. Whereas according to Mayhew (1978, 429ff.) *Trogontherium* is sooner to be looked upon as an inhabitant of stagnant to slow-moving waters, *Castor* may at least partially have been excluded from such biotopes in times of sympatry with *Trogontherium*. Only because *Trogontherium* became extinct in the Middle Pleistocene, its preferred habitat, too, was taken over by *Castor*. *Castor fiber* distinguishes itself by building lodges from felled trees and shrubs, from time to time also by constructing dams and channels. Its staple diet is mainly composed by aquatic plants and – predominant in winter – branches and bark of softwood, especially of poplar, alder, and willow. *Trogontherium cuvieri* possibly preferred the presence of largely unforested enclaves along the waters it inhabited. Such enclaves exist during regenerative phases after high flood events (Mayhew 1978, 433).

Special palaeoecological importance should be ascribed to the semiaquatic hippopotamuses (*Hippopotamus amphibius antiquus*), many remains of which have been recovered at Untermassfeld.

[14] Associated with *Hippopotamus* (Jánossy 1962), as at Untermassfeld.

In the daytime, the stationary animals occupy shallow, open, and in part muddy waters, the banks of which are not too densely forested. Stony parts and spots where currents are strong are avoided, if possible (cf. Section 5.2. for the risks run by the young animals). Especially male individuals often keep their territories occupied over longer periods (Eltringham 1999, 51). At night, hippopotamuses seek out areas of low-growing meadows rich in Gramineae. In contrast with the animals' aquatic residences, these are not subdivided into territorial sections (Eltringham 1999, 49). Food resources are also found away from the water, although female parents accompanied by their calves prefer shorter distances (Eltringham 1999, 53). It becomes apparent here that in landscapes abounding in water *Hippopotamus* has an opportunity to migrate without strictly following the courses of rivers. In the case of the upper and middle sections of the Werra valley here considered, the larger part of the grazing land of the hippopotamuses probably lay beyond the floodplain, on the adjacent Bunter sandstone and Muschelkalk plateaus (see below).

Among the land snails – the majority of which were probably borne into the fluviatile sands from locations in close vicinity to the site – the moisture-loving small amber snail (*Succinea oblonga*) predominates. To date, distinctly sylvan forms have been lacking almost completely. Only the craven door snail (*Clausilia dubia*) and the lapidary snail (*Helicigona lapicida*) are represented in the samples by just a few, partly fragmentary shell remains each.

On the other hand, loose stands of deciduous forest with a distinct layer of herbs or with open spaces are to be inferred from the finds of agile frog (*Rana* cf. *dalmatina*) and common toad (*Bufo* cf. *bufo*). The recent agile frog (*Rana dalmatina*) often lives quite remote from waters and in dryer places than tree frog and moor frog. It prefers light beech forests or mixed woodland and shrubs. The fat dormouse (*Glis sackdillingensis*), a nimble climber, was likewise a forest-dweller.

For the hedgehog (*Erinaceus* cf. *europaeus*) dry biotopes in woodland and more open terrain are typical. The common toad (*Bufo bufo*), too, inhabits forests, thickets, and meadows. Whereas the francolin [*Francolinus (Lambrechtia) capeki*] stayed in the sheltering coverts on the riverbank, the song thrush [*Turdus* sp. (aff. *philomelos*)], the raven [*Corvus* sp. (aff. *janossyi*)], and the Eurasian jay (*Garrulus* aff. *glandarius*) nested at different heights, mostly in trees or shrubs. As a typical woodland bird, the Eurasian jay (*Garrulus glandarius*) feeds on insects, fruit, and sometimes also on birds' eggs. For foraging, *Corvus* probably sought out open country as well.

Clear indications of gallery forest biotopes along the Early Pleistocene Werra are furnished by the suids, several rodents, at least some of the recorded cervids, and the macaques. Without any doubt the wild boar (*Sus scrofa priscus*) preferrred woodland biotopes that amply provided cover and food. The pools and marsh areas of the floodplain gave them the opportunity to wallow. The squirrel (*Sciurus* cf. *whitei*), the vole *Pliomys episcopalis*, and the bank vole *Clethrionomys* cf. *hintonianus* likewise provide evidence of tree stands. Humid unconsolidated soils are indicated by the recovered remains of moles (*Talpa* cf. *minor*, *T.* cf. *europaea*) and shrews (*Sorex runtonensis*).

The ecological parameters of *Bison menneri* remain rather insecure. For Flerov (e.g., 1979), lightly-built, long-legged bisons are generally regarded as woodland forms. According to this notion, the large wild bovid mainly inhabited the sparser, dryer gallery forests of the Werra valley and presumably also wooded landscape sections beyond the floodplain of the river (see below). The relatively frequent occurrence of its skeletal remains supports such an interpretation.

By the same standard, *Eucladoceros giulii* also found refuge in the riverine forests. To a lesser degree this may also be true of the smaller *Cervus* s.l. *nestii vallonnetensis*, which is likewise fairly abundant. Hemmer (2001, 726f.) explains the basic importance of the small cervid in terms of the prey range of the large cheetah *Acinonyx pardinensis pleistocaenicus*, which was also recorded from Untermassfeld. *Cervus* s.l. *nestii vallonnetensis* would accordingly have inhabited more open parkland outside the river valley, like recent fallow deer (see below).

Within a single lineage, the Untermassfeld elk, *Alces carnutorum*, connects the late Middle to Late Villafranchian *Alces gallicus* with the early Middle Pleistocene *A. latifrons*. It must therefore be assumed that male animals of the Untermassfeld population had palmate antlers that jutted out, which feature in turn precludes a life in dense forests. *A. gallicus* mainly inhabited temperate steppes. At middle latitudes in Eurasia, *A. latifrons* was likewise at home in steppe-like biotopes, but also in open meadow landscapes (R.-D. Kahlke 1994, 79; Breda et al. 2005). Similar preferences are to be assumed for *Alces carnutorum*. It probably lived on the park-like plateaus at a somewhat greater distance from the site, as well as in more open parts of the Werra floodplain, as far as the zones bordering upon the river.

The position of the roe deer (*Capreolus cusanoides*) in the Untermassfeld ecosystem is more difficult to assess. The animals certainly preferred biotopes that provided much cover, but it remains unclear whether there are purely taphonomical, predominantly ecological, or even biogeographical causes behind the small number of finds from Untermassfeld.

Because of their climbing capacities, the macaques (*Macaca sylvanus* ssp.) (cf. Section 5.2.), too, are seldom represented among the material recovered. By analogy with recent representatives of the species, they probably liked woodlands with many open spaces, little undergrowth, and also – if available – rocky cliffs. Unlike Zapfe (2001, 894), Haltenorth & Diller (1977, 270) emphasized that the animals prefer the presence of water in the vicinity. However, the entirety of the recent genus encompasses more diverse ecotypes: macaques live in forests that differ in altitude and degree of humidity as well as in open grassland, and also in bush and cactaceous steppes (Franzen 1973, 355). Parts of the Untermassfeld population may therefore also have found suitable living conditions beyond the river valley.

The woody and watery biotopes of the Early Pleistocene floodplain of the Werra formed the main living space of the jaguar (*Panthera onca gombaszoegensis*). Its build is comparable to that of the recent jaguar and allows the inference that its affinity for water was similar. Such a conception is also corroborated by the circumstances found for other old-world fossil occurrences of the jaguar (Hemmer et al. 2001). According to Hemmer (2001, 710), the prey range of the fossil form from Untermassfeld included *Cervus* s.l. *nestii vallonnetensis*, *Eucladoceros giulii*, at least the calves of *Bison menneri*, and in addition now and then *Capreolus cusanoides*, *Alces carnutorum*, and *Sus scrofa priscus*. It seems possible that calves of elephants, rhinoceroses, and hippopotamuses were also attacked. *Trogontherium* and *Castor* may be looked upon as attainable small additional prey.

Ursus cf. *dolinensis* from Untermassfeld shows many primitive characteristics, which it shares with the brown bears (Musil 2001b; Garcia 2004). Hence the ecological parameters of today's brown bears are in principle likely to be transferable to the bear from Untermassfeld. The geographically and seasonally varying omnivorous dietary pattern accordingly included an essential vegetable component. On the one hand, the carnivorous section of the food range was covered by hunting for prey species up to the size of *Eucladoceros* and *Alces*, but on the other hand *Ursus* cf. *dolinensis* may be looked upon as an opportunistic scavenger comparable to recent brown bears. During periods of activity, the animals certainly roamed through areas of considerable extent, so that there was no direct connection with special biotopes or forest types, but rather with the entirety of biotopes for foraging, daybeds, transit regions, and winter dens (cf. Heptner et al. 1974, 430).

Clues for the ecological character of various locations on the slopes near the site, especially of the south- to southwest-facing right slope of the valley (cf. Fig. 77), may in particular be deduced from the gastropod remains transported into the site. The relatively poor snail fauna (cf. Section 3.2.) from Untermassfeld includes comparatively many inhabitants of open biotopes. Beside the small amber snail already mentioned, the hypogeal blind awl snail (*Cecilioides acicula*) occurs more frequently. It prefers dry and warm habitats, i.e., dry grassland (with some shrubs locally) and stone fields, as they are typical of, especially, south-facing locations on the Muschelkalk slopes. Through the activities of animals that turn up the soil, but particularly as a result of heavy precipitation, the shells are brought to the surface (Zeissler 1981, 41). Moss snail (*Pupilla muscorum*) and smooth grass snail (*Vallonia pulchella*), whose remains are likewise fairly numerous in the samples, are mostly found in open terrain, but also in shrub vegetation and light woodland (Zeissler 1981, 21-22). In addition, *Helicigona lapicida* is indicative of rocky landscape components. The gastropod remains from Untermassfeld, although they are often fragmentary, yet for the most part show a relatively good material preservation, so that it has to be assumed that the majority of the specimens were washed in from the immediate surroundings.

Certain locations in the Werra valley, especially on the south-facing slope with its cliffs and other opportunities for taking cover, were very eligible for badger (*Meles hollitzeri*) and lynx (*Lynx issiodorensis* ssp. ex aff. *spelaeus*) to establish breeding lairs and to raise their cubs. With reference to the recent species *L. rufus* and *L. spelaeus*, which are nearest to it in their build, *Lynx issiodorensis* is to be credited with a great ecological plasticity (Hemmer 2001, 739f.). If there were opportunities to take cover, the biotopes that contributed to its food supply probably extended from the floodplain along the river onto the adjacent plateaus. Roe deer (*Capreolus cusanoides*) and hare (Leporidae) most likely were its main prey. In addition, fawns of *Cervus* s.l. *nestii vallonnetensis*, young of *Sus scrofa priscus* and – as far as attainable – also the two beaver species may have been hunted. According to their body sizes, macaques, badger, small rodents and birds also fit into the prey range of the Untermassfeld lynx (Hemmer l.c., 740).

Beyond the upper and middle Werra valley lay plateaus where Muschelkalk or Bunter sandstone subcropped. These plateaus drained into the river. Sparsely wooded to open biotopes in the wider surroundings of the site are indicated by several elements of the Untermassfeld fauna. To begin with, *Mammuthus* and *Stephanorhinus* are two large mammals that were suited to inhabit savanna-like landscapes as well as sparse woodlands. With the phylogenetic transition from late southern mammoths (*Mammuthus meridionalis* ssp.) to early steppe mammoths (*M. trogontherii trogontherii*), the ecological range of tolerance within the Eurasian mammoth lineage shifted from habitats within temperate tree savannas to steppic landscapes of a more continental character (R.-D. Kahlke 1994, 76). As shown by Garutt & Bajguševa (1981, 13-15), their high and relatively short build enabled the southern mammoths to get hold of leaves, branches, and bark. In principle this is also applicable to the very high stature of the mammoths from Untermassfeld (body height calculations in Dubrovo 2001, 603). It seems likely that the food range of the Untermassfeld mammoths was subject to a marked seasonal change. Whereas fresh vegetation from open country was preferred in spring and at the beginning of summer, forage from trees – certainly also from those of the forests in the floodplain – became more important later in the year.

Stephanorhinus etruscus probably had similar preferences, although it exploited a different storey of the vegetation. When compared with older or more southern populations, the rhinoceroses from Untermassfeld bear evidence of a lengthening of the extremities (H.-D. Kahlke 2001b, 549, as well as the Figs. 53-56 of the present contribution). This feature marks the beginning of an adaptational process that later allowed the animals to endure increasing continentality, at least in Central Europe (e.g., the early Middle Pleistocene of Süssenborn).

Characteristic inhabitants of more open parts of the landscape beyond the river valley were the green toad (*Bufo* cf. *viridis*), as well as a frequently demonstrated ground squirrel [*Spermophilus (Urocitellus)* ex gr. *primigenius / polonicus*], a hamster (*Cricetus* cf. *runtonensis*), a vole (*Microtus thenii*), a hare (cf. *Lepus* sp.), and among the large mammals especially horse (*Equus wuesti*) and cheetah (*Acinonyx pardinensis pleistocaenicus*).

The recent green toad (*Bufo viridis*) avoids woodland areas. It tolerates aridity and certain seasonal temperature fluctuations, which allows an advance into steppe regions. *Equus wuesti*, an equid close to the small *E. altidens* (Musil 2001a), has to be looked upon as a more or less stenotopic grazer that found agreeable living conditions in the open plains along the Werra valley (see Forstén 1990, 71 on the ecology of smaller stenonid equids).

In the construction of its body, the Untermassfeld cheetah shows the same typical features as the recent species *Acinonyx jubatus*, which allows the inference that the fossil form was a hunter in open landscapes and developed superior speeds on short and medium distances (Hemmer 2001, 725ff.). The animal therefore avoided dense tree and shrub stands as much as entirely open terrain that offered no possibilities for taking cover. Its preferred habitat was the park landscape. It is true that the great body mass of the large fossil cheetah did not allow the high sprint velocities of the recent species, but on the other hand the hunter in the Untermassfeld landscape was dealing with correspondingly heavier prey species. According to Hemmer (l.c., 727) these were especially the small *Cervus* s.l. *nestii vallonnetensis* and, in the appropriate season, foals of *Equus wuesti*. Insofar as they stayed in the open country, calves of *Bison*, *Eucladoceros*, and *Alces* probably also belonged to the prey range of *Acinonyx*. Once in a while the small wolf (*Canis mosbachensis*) may have been seized as well.

The group of Untermassfeld carnivores that were linked with a wide variety of habitat types or had no recognizable connections with special environments comprises the above-mentioned wolf (*Canis mosbachensis*) and also the puma (*Puma pardoides*), a larger canid [*Canis (Xenocyon) lycaonoides*], the dirktooth cat (*Megantereon cultridens adroveri*), a large hyaena (*Pachycrocuta brevirostris*), and the sabretooth cat (*Homotherium crenatidens*).

By analogy with the prey range of recent (smaller) southern wolves, *Canis mosbachensis* would have been able to kill animals the size of *Cervus* s.l. *nestii vallonnetensis* at the very most. However, small animals such as hares, rodents, and birds were probably seized more often (Hemmer l.c., 768). In addition, carcass remains from kills made by larger carnivores must be taken into account as an important food source. The apparently high population density of the small wolves (cf. Section 5.2.) could have led to a dominant position with respect to the foxes, whose presence has not been substantiated by the material recovered to date.

To judge by the pelvic and tibial remains at hand, the jumping performance of the puma recorded from Untermassfeld (*Puma pardoides*) was inferior to that of the recent form (*Puma concolor*) (Hemmer l.c., 732). According to Hemmer (l.c., 734), its prey range included ungulates of intermediate size, viz. *Cervus* s.l. *nestii vallonnetensis*, at least relatively young or small boars (*Sus scrofa priscus*), and roe deer (*Capreolus cusanoides*). Intermittent attacks on larger cervids (*Eucladoceros giulii*, calves of *Alces carnutorum*) must be allowed for. Hares, the two beaver species (*Castor fiber, Trogontherium cuvieri*), and macaques (*Macaca sylvanus* ssp.) are in the lower section of the prey range.

Hemmer (l.c., 767) supposed that socialization in *Canis (Xenocyon) lycaonoides* was highly developed. If it hunted in packs, *Canis (Xenocyon)* would have to be looked upon as the top carnivore of the Untermassfeld fauna. Its wide prey range could then have included rhinoceroses (*Stephanorhinus etruscus*), young hippopotamuses (*Hippopotamus amphibius antiquus*) as well as young mammoths (*Mammuthus*).

The dirktooth cat (*Megantereon cultridens adroveri*), too, attacked large herbivores. These could sooner be overcome in a surprise attack and their carcasses could for a longer period serve as food reserves (Hemmer l.c., 748). Because the upper canines were high-crowned and therefore in danger of breaking, bites that were meant to kill were probably directed towards the region of the throat. Bison (*Bison menneri*), rhinoceros (*Stephanorhinus etruscus*), and presumably also young mammoths (*Mammuthus*) are likely to have been attainable prey. If an aggressive takeover of prey was attempted (by, e.g., *Panthera onca gombaszoegensis*), the strong warning function of the canines was brought into play (Hemmer l.c., 748ff.).

Pachycrocuta brevirostris should be interpreted as an opportunistic scavenger, which by its impressive body size was likewise in a position to coerce other carnivores into relinquishing their prey (Turner & Antón 1996, 463-464). It is highly probable that territorial marking (Keiler 2001c, 696) characterizes the Untermassfeld site as an at least temporarily very attractive part of a hyaena territory. If carcasses became available, the animals must also have extended their activities into areas of the Werra valley floodplain that had been scoured level by the floods. Large-scale transport of carcass parts in the form of material brought in by *Pachycrocuta* was documented at, e.g., Zhoukoudian, Loc. 1 (Pei 1934, 90ff.). With regard to Venta Micena, selective transport of carcass parts is a subject of discussion (Arribas & Palmqvist 1998, 40). Where the taphonomical ensemble of Untermassfeld is concerned, it must be assumed that material was carried off the site (cf. Sections 4.6, 5.4).

The activities of the large sabretooth cat (*Homotherium crenatidens*) did not show a recognizable habitat specialization either. To judge by the build of its body and limbs, it was specialized in rapidly traversing extensive areas (Hemmer 2001, 758). The display function of its powerful upper canines in combination with its body size probably guaranteed *Homotherium* the leading position within the carnivore hierarchy (see below). The structural specialization of its physique, which shows parallels with bears and hyaenas, allows the conclusion that it availed itself of carrion and also actively hunted (Hemmer l.c.). It must be assumed that its killing technique, which it had in common with *Megantereon*, was a bite in the throat region of large and fleshy prey. In addition, the long upper canines were probably advantageous when large carcasses had to be opened (e.g., of *Mammuthus*, *Hippopotamus*, and *Stephanorhinus*). Apart from prey stolen from other carnivores, calves of elephants, and now and then hippopotamuses and rhinoceroses, were likely hunting targets (Hemmer l.c.).

On the basis of reconstructions of habitats and specializations in killing techniques or prey, Hemmer (l.c., 762ff.) discussed the aptitude for coexistence of the six felids represented in the Untermassfeld fauna (or seven, if the European wild cat, whose presence is very much likely, is included). If all average- to large-sized carnivores from the Untermassfeld site are considered with regard to their potential for killing or stealing prey, the following interspecific hierarchy presents itself: *Canis (Xenocyon) lycaonoides* (if it formed packs) --> *Homotherium crenatidens* --> *Megantereon cultridens adroveri* --> *Panthera onca gombaszoegensis* --> <-- *Ursus* cf. *dolinensis* --> *Pachycrocuta brevirostris* --> *Acinonyx pardinensis pleistocaenicus* --> *Canis (Xenocyon) lycaonoides* (as solitary hunters) --> *Puma pardoides* --> *Canis mosbachensis* --> <-- *Lynx issiodorensis* ssp. A synecological chart showing probable interrelations between the carnivores as well as their predator-prey relations was worked out by Hemmer (l.c., fig. 20).

6.2. Landscape reconstruction

The entirety of the available results of geological, taphonomical, and palaeozoological-ecological research allows a fairly detailed description of the landscape that existed at the time the complex site of Untermassfeld came into being.

When the water level was normal, a dynamic stream passed through the Early Pleistocene Werra valley, which at some intervals was rather wide and flat-bottomed but, in keeping with its geological age, was still shallow. The bed as well as the banks of the river consisted of multicoloured gravels, reddish sands of variable grain sizes, and a variety of finely clastic floodplain sediments. Here and there the bottom of the water was covered by a lush vegetation. In protected sections of the river there were open and clear expanses of water where currents were weak. Within parts of the floodplain that were saturated with water, next to the main stream or its more or less active lateral channels, there were stagnant blind arms and zones that were silting up. Special meteorological-hydrological constellations could cause a surge of the water volume flowing, and change the Werra into a powerful stream, or into a swirling torrent where the valley became narrower (at the „Needle's Eye" for example, 12 km upstream of the site). In the floodplain the water was regularly pushed up to abnormal heights.

Because of their deviant water supply, riverine floodplains normally contain biotopes with characteristics that differ from the zonal ecological relations in the region (Walter & Breckle 1986, 124ff.). Temporary ponds, pools, and puddles, distributed according to the impermanent floodplain relief, existed in the Early Pleistocene Werra valley, probably also as a legacy of regularly occurring high flood events. Depending on insolation and depth, the temperatures in these little waters were diverse. Humid water meadows had partly turned into swamps. Groundwater outflows from the then youngest gravel terrace could also have contributed to their permanent watering. A rather large number of permanently humid locations was probably covered with softwood stands, so that for instance the presence of waterside thickets in close vicinity to the river may be taken into account. In relation to the duration of floods and the intervals between them, trees in areas particularly endangered by high water were more or less damaged and hence largely low-growing. Sparse mixed woodland with tall trees and less undergrowth extended across more stable and relatively dry locations in the floodplain. A lush herbaceous vegetation grew on its often loose soils. All in all, the variably distributed wet, humid, and relatively dry locations in the Werra floodplain offered suitable living conditions to a wide variety of plant communities.

Beside biotopes with woodland and shrub vegetation poor in species, the Muschelkalk slopes near the site also included open areas. Dry grassland and barren scree fields were particularly common in south-facing places on the right valley slope (where the site is). Rock characteristics suggest limestone banks bared by erosion (so-called "Werksteinbänke" = freestone banks) and crumbling cliffs (Wellenkalk). Clayey / limy talus cones had formed at the foot of the slope.

The plateaus that flanked the Werra valley were covered by loamy / clayey soils that were rich in limestone debris and overlay Muschelkalk outcrops (Middle Triassic), as well as by sandy / loamy soils that overlay Bunter sandstone (Lower Triassic). The surfaces of the Lower Muschelkalk, particularly on top, tended to dry out rapidly. Outside the Werra valley, park-like landscapes and sparsely wooded to open biotopes, according to the soil characteristics, extended over large distances.

6.3. Climatic parameters

The composition of the Untermassfeld fauna, and the landscape reconstruction based on it, undoubtedly indicate warm climatic conditions. The fossils recovered mainly represent temperate and thermophilous elements. Animals characteristic of cool- to cold-stage conditions are totally absent.

A skink (*Chalcides* sp.) of Mediterranean affinity and – as an exotic species of Asian origin – a gecko (*Eublepharis* sp.) must, for example, be looked upon as faunistic indicators of warmth. *Macaca*, too, is mainly found in fully interglacial faunal communities of the Central-European Pleistocene (compilations in Delson 1980, 19; Koenigswald 1991, 81). However, occurrences under boreal conditions cannot be totally precluded here (Franzen 1973, 355; Fladerer 1991, 281). A certain link with warmth may also be deduced from the predominantly northern Mediterranean distribution of the small cervids related to *Cervus* s.l. *nestii vallonnetensis*. Something similar goes for *Francolinus capeki*, osteologically the closest relative of the recent francolin (*Francolinus francolinus*), nowadays distributed from Asia Minor to Middle Asia. Finally, *Hystrix* was probably bound to temperate biotopes (Mohr 1965).

Higher summer temperatures than occur in the southern Thuringian region today, however, are indicated by the find of a freshwater turtle (Emydini). A successful incubation of embryos of this group can only take place if there are well over 60 summer days with air temperatures rising above 25°C (Maul 1997, 92-93; principle data in Stuart 1979). The winter temperatures in the Untermassfeld area at the time the site came into being may be estimated by the frequent occurrence of *Hippopotamus amphibius antiquus*. Because survival of the hippopotamuses may be ruled out if their watery stations would have been covered by ice, mild winters with minimum air temperatures of a few degrees below zero may probably be inferred. Regular and lasting snow covers during winter are therefore rather unlikely for the landscapes in close vicinity to the site.

From the determination of $^{18}O/^{16}O$ ratios in the phosphate groups of bone apatite from horse metapodials, Stephan et al. (2001) calculated yearly temperature averages for the area around Untermassfeld. These lie approximately 1 degree above the present average temperatures in the region. Moreover, the diagenetic apatite recrystallization demonstrated for the analysed bone material could have led to a secondary decrease of the $\delta^{18}O$ values of newly-formed phosphate groups, and hence to a lowering of the estimated temperature values (Stephan et al. 2001, 926). In addition, the relatively nearby low mountain ranges (Thuringian Forest and Thuringian Slate Mountains), with their higher levels in part directly draining into the Werra (cf. Section 5.1. and Fig. 78), could to a certain extent have cooled the river water, which was an essential reservoir for watering the large mammals. One may suspect that this, too, was a cause of at least minor $\delta^{18}0$ value reductions in the bone apatites of the Untermassfeld equids. The yearly temperature averages calculated to have been 1 degree above the present values must therefore be considered as minimum values.

In keeping with variable humidity and insolation, various local climates may be reconstructed for the riverine floodplain, locations on slopes, and plateaus in the wider surroundings of the site. The regional precipitation sufficed to maintain open expanses of water and a very humid floodplain with variegated vegetation in the Werra valley throughout the year. Especially during the winter half-years the precipitation became heavy and repeatedly led to high flood situations. The permanent draining of the adjacent plateaus contributed to the saturation of the floodplain all through the year, so that assuming similar air temperatures, the conditions on the plateaus were generally dryer than in the valley. The Muschelkalk slopes could only retain small quantities of water. In south-facing places the dryness was combined with a greater than average warming of the soil and the air, particularly during the summer months.

Regionally, a temperate warm-humid climate prevailed, balanced by Atlantic influence and free from extreme daily or seasonal fluctuations in temperature.

7. Stratigraphic age[15]

7.1. Litho- and magnetostratigraphy

If one follows the lithostratigraphical classification of the gravel terraces of the middle Werra valley according to Ellenberg (1968), the Younger Weathered Coarse Gravels underlying the Untermassfeld site were accumulated during the Eburonian. In Waalian times, a profound chemical weathering of the felspar components in the boulders and pebbles of both volcanic rock from the Rotliegendes and of sandstone followed, with illite and montmorillonite as end products (Ellenberg & R.-D. Kahlke 1997, 58-59; Ellenberg et al. 1997, 36-38, fig. 1). Unger (in Unger & R.-D. Kahlke 1995, 204-206, fig. 4A) thought the gravel deposition could only have come to an end as late as the Menapian (cf. also Ellenberg 2002, 241ff.). In this case the weathering would have taken place at an accordingly later date, i.e., during predominantly warm phases before the normally magnetized part of the Bavelian (according to Zagwijn & Jong 1984, 166-167, the basis of the Jaramillo lies within the Bavel Interglacial).

The (unweathered) sequence of layers (floodplain sediment to Upper Fluviatile Sands) cut by the Untermassfeld excavation postdates the weathering process documented in the underlying gravels. In accordance with the above-mentioned basic geological data it may be stated that the deposition of the fossiliferous sands of Untermassfeld – which no doubt took place within a relatively short time – possibly had already occurred during the late Waalian at the earliest, but more likely either during the Bavelian or an early part of the Cromerian Complex (Cromerian s.l.) (R.-D. Kahlke 1987, 24; Ellenberg & R.-D. Kahlke 1997). The preservation of the Sands was effectuated by neotectonic tilting of a ridge-like horst southwest of the Marisfeld fault zone, and a concomitant subsidence of both the Younger Weathered Coarse Gravels and the overlying Fluviatile Sands to a level approximately 10 m below the erosional plane of the Werra (Ellenberg 1995, 419; Ellenberg et al. 1997, 36). At lower valley levels of the middle Werra valley the next younger gravel bodies found are Cromerian and subsequently Elsterian terrace remains (Ellenberg & R.-D. Kahlke 1997, table 2, fig. 3).

The reversed magnetization of the entire underlying Quaternary sequence within the area of the site (Younger Weathered Coarse Gravels to middle section of the Lower Fluviatile Sands), as well as the normal magnetization of both the upper section of the Lower Fluviatile Sands (high flood sediments) and the Upper Fluviatile Sands (Channel Infill), offer two possibilities for a litho-magnetostratigraphical positioning of the Untermassfeld fauna. It correlates either with an early section of the Early Pleistocene Jaramillo polarity subzone (Jaramillo event), or with the beginning of the Brunhes polarity zone (Brunhes magnetochron). With respect to Untermassfeld, the likewise normally magnetized Cobb Mountain cryptochron may be excluded because of its absolute age of approximately 1.2 Ma BP (Cande & Kent 1995, table 3.; cf. Section 7.2.1.)

7.2. Biostratigraphy

Examination of the palaeozoological finds from Untermassfeld should lead to a more precise statement regarding the stratigraphical interpretation of the fossiliferous sands, which rests on analyses of geological and palaeomagnetical data. At the same time, the biostratigraphical information contained in each of the animal groups is rather diverse. As expected, the malacofauna (Krolopp 1997), ichthyofauna (Rutte 1997), amphibian (Böhme 1997) and turtle remains (Maul 1997), as well as the avifauna (Jánossy 1997), but also some mammalian taxa, e.g., *Macaca sylvanus* (Zapfe 2001), rather allow statements on biodiversity, ecological and biogeographical conditions, as well as on processes relating to taphonomy and genesis, than on the stratigraphical position of the fossiliferous deposit.

7.2.1. Maximum age

The existence of a clear chronological dividing line between the mammalian association from Untermassfeld and the older faunas of the European Villafranchian becomes apparent if one considers the following taxa:

1. The tall and slender *Bison menneri* has replaced the smaller and slighter *Leptobos* forms known from Villafranchian sites (Les Ètouaires, Dusino, Villaroya, Saint-Vallier, Le Coupet, Tegelen, Olivola, Senèze, Peyrolles, Tasso, Farneta) and also has substituted the representatives of

[15] In the following discussion the (provisional) boundary between Early and Middle Pleistocene – in the sense of the XIIth INQUA Congress (Ottawa 1987) and in acceptance of the concurring arguments by Richmond (1996) – is equated with the Matuyama / Brunhes boundary (cf. also R.-D. Kahlke 1997b, footnote 4).

the *Eobison* group (Pirro Nord, Venta Micena) (Sher 1997).

2. *Capreolus cusanoides* has so far not been demonstrated in older associations, although the scarce remains of roe deer, or generally of Middle Villafranchian to early post-Villafranchian cervids of comparable size (Beremend 5, Villàny 3, Matassino, Casa Frata, Valea Grăunceanului, Pirro Nord), do not allow a decisive comparison.

3. *Alces carnutorum* from Untermassfeld surpasses the Middle to Late Villafranchian *A. gallicus* (from Senèze for instance) in size, and hence is transitional between the latter and *A. latifrons* (H.-D. Kahlke 1995a, 1997). The finds from Untermassfeld mark the northeastern marginal zone of the distribution area of European Villafranchian elks as it has been known until now (H.-D. Kahlke 1990b, 85; Breda & Marchetti 2005, fig. 3).

4. As a late representative of the *rhenanus* / *nestii* group, *Cervus* s.l. *nestii vallonnetensis* from Untermassfeld continues the late Middle to post-Villafranchian trend towards a relative shortening of the premolar rows (Olivola, Casa Frata, Pirro Nord) (H.-D. Kahlke 1997).

5. A tendency in *Eucladoceros* to develop particularly large forms (H.-D. Kahlke 1997) becomes visible during the period from the Late Villafranchian to the early post-Villafranchian (Selvella, Sainzelles, Pirro Nord) and finds full expression in *E. giulii*, which is equipped with, among other things, extraordinarily long metapodials.

6. To judge by the cranial characteristics in particular, *Hippopotamus amphibius antiquus* from Untermassfeld represents a more evolved form with respect to the Late Villafranchian finds (Figline, Upper Valdarno) (R.-D. Kahlke 2001e).

7. *Sus scrofa priscus* first appeared (Guérin & Faure 1997) in the Epivillafranchian (Ceyssaguet) and in Europe replaced the Villafranchian *Sus strozzii* (Valdeganga 2, Le Coupet, Mugello, Tegelen, Olivola, Senèze, Quercia, Upper Valdarno).

8. Whereas they are to a great extent morphologically and metrically close to late Middle to Late Villafranchian populations (Olivola, Pietrafitta), the Untermassfeld finds of *Stephanorhinus etruscus* show signs of a lengthening of the extremities, which is even more strongly expressed in early Middle Pleistocene forms known from northerly occurrences (H.-D. Kahlke 2001b).

9. The horse remains from Untermassfeld, described as a new species *Equus wuesti* by Musil (2001a), both morphologically and metrically differ from the finds from Middle Villafranchian to early post-Villafranchian occurrences (Chilhac, Olivola, Tasso, Farneta, Sainzelles, Pirro Nord, Venta Micena). Musil therefore considers a correspondingly younger age to be possible. Nevertheless, Forstén (1998) pointed out the limited suitability of *Equus* species as stratigraphical guide fossils.

10. The elephantid remains from Untermassfeld by no means represent the typical Middle to Late Villafranchian *Mammuthus meridionalis meridionalis* (Tegelen, Olivola, Tasso, Dmanisi). Dubrovo (2001) referred the material to *Mammuthus trogontherii trogontherii*. However, as was ascertained by Lister (1996a, 207-208), not all (late) *meridionalis* mammoths possess a serial carpal build that differs from the condition found at Untermassfeld. Also, because of the still very small number of available remains of the dentition, the presence of an advanced form of *Mammuthus meridionalis* at Untermassfeld cannot be excluded (Lister et al. 2005, 54)[16].

11. On both morphological and metrical grounds, the Untermassfeld jaguar must be referred to the post-Villafranchian / early Middle Pleistocene chronosubspecies *Panthera onca gombaszoegensis*, not to the late Middle to Late Villafranchian *P. onca toscana* known from Tuscany (Olivola, Perugia, Upper Valdarno) (Hemmer 2001).

12. The large cheetah from Untermassfeld morphologically differs from the forms of the European Villafranchian (Les Étouaires, Montopoli, Villaroya, Saint-Vallier, Olivola, Casa Frata) and should be attributed to *Acinonyx pardinensis pleistocaenicus*, which has so far only been described from northern China (Hemmer 2001).

13. It is true that the Untermassfeld lynx shows affinities with the Late Villafranchian (Tasso) forms of southern Europe (*Lynx issiodorensis valdarnensis*), but in some respects it already approaches that of *Lynx spelaea*; overall, it is intermediate (Hemmer 2001).

14. Together with the finds from Venta Micena,

[16] In Europe and Asia many evolved local forms of *M. meridionalis* were named at the subspecific level without having been extensively compared with one another (R.-D. Kahlke 1994, 25-26; Lister 1996, 207; Lister et al. 2005, 6), which renders a more precise stratigraphical statement difficult.

the dirktooth cat from Untermassfeld has been separated from typically Villafranchian representatives of the species (Saint-Vallier, Pardines, Olivola, Senèze, Figline) and referred to *Megantereon cultridens adroveri* [Hemmer 2001; see also Sardella (1998) on the taxonomical interpretation of Old-World *Megantereon* remains].

15. The Untermassfeld bear *Ursus* cf. *dolinensis* shows a series of primitive characteristics, which it shares with the older populations of *U. etruscus* (Saint-Vallier, Kuruksai, Senèze, Tegelen, Olivola, Upper Valdarno, Pietrafitta; Mazza & Rustioni 1992). The larger dimensions of its body and dentition (Musil 2001b) as well as the incipient reduction of the premolars (Garcia 2004) make it stand out against *U. etruscus* as a younger evolutionary stage.

16. The large canid *Canis (Xenocyon) lycaonoides* from Untermassfeld, as well as the post-Villafranchian forms known from the Cueva Victoria and Venta Micena, differ from typically Late Villafranchian forms [*C. (Xenocyon)* ex gr. *falconeri*] by a reduced or absent entoconid on the M_1 as well as by a series of other characteristics of the dentition (Sotnikova 2001).

17. The proportions of skull and dentition, and also some dental characteristics separate the small Untermassfeld canid – as *Canis mosbachensis* (or *C. lupus mosbachensis*) – from late Middle Villafranchian to earlier post-Villafranchian populations (Olivola, Tasso, Pirro Nord, Cueva Victoria, Venta Micena) (Sotnikova 2001).

18. To judge by the evolutionary level of *Microtus* and *Mimomys savini*, and in part also by the evolutionary stages of *Drepanosorex*, *Spermophilus primigenius / polonicus*, and *Glis sackdillingensis*, the small mammal fauna from Untermassfeld is younger than well-researched faunas of the older Early Pleistocene (Soave / Cava Sud, Pietrafitta, Osztramos 8, Pirro Nord 1, Venta Micena 1 and 2, Betfia II, Deutsch-Altenburg 2) (Maul et al. 1998; Maul 2001a).

Most large mammal taxa present at Untermassfeld quite clearly separate its fauna from Villafranchian associations. Moreover, some of the forms (*Bison menneri*, *Cervus* s.l. *nestii vallonnetensis*, *Eucladoceros giulii*, *Ursus* cf. *dolinensis*, *Canis mosbachensis*) suggest a younger date with respect to comparatively early post-Villafranchian faunas (Pirro Nord, Cueva Victoria, Venta Micena). According to Masini et al. (1994, 385), the fossiliferous sediments of Pirro Nord show reversed magnetization and are to be considered as older than the Jaramillo polarity subzone. They are with reservation correlated with the Menapian. A similar age of around 1.2 Ma BP was made probable for Venta Micena on the basis of faunistic data (Moyà-Solà 1987, 222; Agusti & Moyà-Solà 1991, 757ff., 1992, 72). Absolute dates obtained from amino-acid racemization (Torres et al. 1997) are widely scattered here (591 $\pm$ 77 to 1511 $\pm$ 318 ka). Whereas Aguirre & Morales (1990, 9) considered the karst infill of the Cueva Victoria to be of similar age to the faunal community of Venta Micena, Agusti (1991, 134, 136) and Agusti & Moyà-Solà (1991, 756ff.) regard the former site as somewhat older than the latter (approx. 1.3 Ma BP).

From these ordinations and from comparisons with other small mammal faunas (Maul 2001a, see above), a maximum age of $<$ 1.2 Ma BP for the fossiliferous sands of Untermassfeld may for the present be deduced.

7.2.2. Minimum age

For the determination of the minimum age of Untermassfeld the following faunistic points present themselves:

1. With some reservation, the presence of *Bison menneri* could indicate a late Early Pleistocene (minimum) age of the site, because bovines that admittedly still shared some traits with *B. menneri* but are rather referable to *B. voigtstedtensis* already occurred, at least in the southwestern European area, at a time close to the Matuyama / Brunhes (=M/B) boundary (Atapuerca ?TD3, TD6, TD8a) (Soto 1987, 99ff.; Made 1998, 278, 1999, 401ff.). During the further course of the early Middle Pleistocene (Durfort, Isernia la Pineta, Süssenborn, Mauer, Mosbach 2), the likewise fairly slender *Bison schoetensacki* and also early representatives of the more robust *B. priscus* or, in eastern parts (Tiraspol) a strong (? priscoid) *Bison* sp., spread over large areas of Europe (Sala 1987, 1997; Brugal 1995).

2. In faunas with a marked open-ground component, the next younger European roe deer finds – after *Capreolus cusanoides* from Untermassfeld – belong to populations of the early Middle Pleistocene *C. suessenbornensis* (Voigtstedt, Süssenborn, Stránská Skála, Miesenheim 1, Würzburg-Schalksberg, Mosbach 2, Bilshausen, etc.) (Mäuser 1987, 33; E. Turner 1990, 304ff.; Pfeiffer 1998; H.-D. Kahlke 2001a). Smaller early Middle Pleistocene roe deer from Mauer have for the time being been referred to *C. priscus* (Pfeiffer 1998, 65).

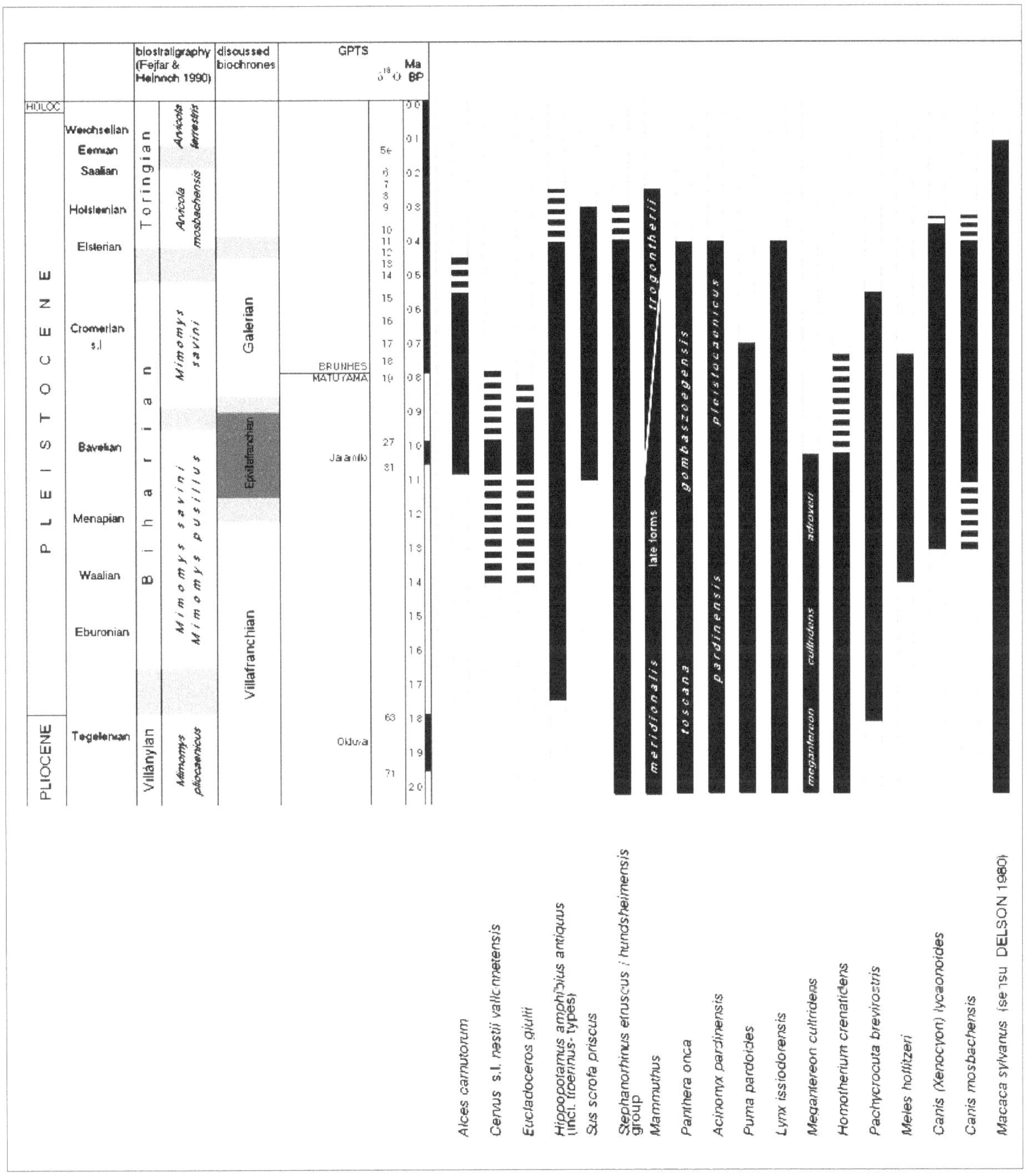

Fig. 79[17,18] European Pleistocene biostratigraphical ranges of the large mammal taxa demonstrated at Untermassfeld.
Compilation with the benefit of information provided by Faure & Guérin (1984); Jánossy (1986); H.-D. Kahlke (1990, 1997); Fladerer (1991); A. Turner (1992a, 1994, 1995); Cerdeño (1993); Fortelius et al. (1993); Wolsan (1993); R.-D. Kahlke (1994); Masini et al. (1994); Rook & Torre (1996); A. Turner & Antón (1996); Alberdi et al. (1997); Guérin & Faure (1997); Sardella et al. (1998); Garcia & Arsuaga (1999); Koenigswald & Heinrich (1999); Hemmer (2001); Sotnikova (2001); Breda & Marchetti (2005), and kindly communicated orally by H. v. Essen, J. v. d. Made, and D. Mol.

[17] New species descriptions based on material from Untermassfeld were only taken into account if their remains also occur at other European sites.

[18] The biostratigraphical ranges of accompanying small mammal taxa are to be found in Maul (2001a, fig. 54).

3. In Central Europe, elk remains different from *Alces latifrons* and attributed to the still poorly known form *Alces carnutorum,* are demonstrable up to a point above the M/B boundary in the *Mimomys savini* fauna of Voigtstedt; and probably still occurred at an even later date in western Europe (Vergranne) (Geraads 1983, 69-72; H.-D. Kahlke 1990b, 86).

4. On the basis of their dimensions, some remains of small cervids from Atapuerca (TD4, TD6, TD8a), chronologically close to the M/B boundary, probably belong to the form *Cervus* s.l. *nestii vallonnetensis* from Untermassfeld rather than to later Middle Pleistocene fallow deer (Made 1998, 273-274, 1999, 395ff.).

According to Made (1998, 276, 1999, 399f.) an elaphine deer is already found below the M/B boundary at Atapuerca (TD4, TD6), in contrast with Untermassfeld.

5. After Made (1998, 275, 1999, 397) had succeeded in making the occurrence of *Eucladoceros giulii* plausible for Apollonia 1 (Kostopoulos 1997, 846ff.), the same author considered the large *Eucladoceros* form to be demonstrable below the M/B boundary (TD4, TD6) at Atapuerca as well, and with a fair degree of confidence.

6. The skull characteristics of *Hippopotamus amphibius antiquus* from Untermassfeld turn out to be less advanced than the (?) late Early Pleistocene find from S. Oreste (Galerian above the Jaramillo), and in particular the skull from Maglianella (Galerian around 0.7 Ma BP) (Mazza 1991; R.-D. Kahlke 2001e). However, the reputedly Galerian *Hippopotamus* skull from Tor di Quinto, compared to the specimen from Untermassfeld, appears to be less specialized (R.-D. Kahlke 2001e). Meanwhile Mazza (1995, 73) considered the stratigraphical position of this piece doubtful. The Untermassfeld specimens differ from younger series of finds from the early Middle Pleistocene (Mosbach 2, Jockgrim, etc.) mainly in their greater robustness and size (R.-D. Kahlke 1997b).

7. In Europe *Sus scrofa priscus* (including *S. scrofa mosbachensis*) persisted through the Middle Pleistocene until the "Mindel / Riss" (Lunel Viel) (Guérin & Faure 1997). Hence a more precise statement concerning the stratigraphical position of the Untermassfeld fauna cannot be based on the suid remains.

8. Populations of the *Stephanorhinus etruscus / hundsheimensis* group continued to exist throughout Europe until some point in the early Middle Pleistocene (e.g., Mosbach 2) (Fortelius et al. 1993). Body size and proportions of *S. etruscus* from Untermassfeld are indicative of a lower stratigraphical position when contrasted with the neighbouring forms from Voigtstedt and Süssenborn, as well as in comparison with those from Hundsheim (H.-D. Kahlke 2001b).

9. Regardless of the referral of the Untermassfeld *Mammuthus* remains to either an evolved *meridionalis* or an early form of *trogontherii*, a more exact stratigraphical statement can hardly be based on these finds. Possibly the earliest evidence of *Mammuthus trogontherii* in Central and western Europe is known from Dorn-Dürkheim 3 (c. 800 Ka, Franzen et al. 2000) and from the somewhat younger English Beestonian [Beeston cliff, Lister et al. 2005]. On the other hand, remains of very late, evolved *meridionalis* populations were reported from Dorn-Dürkheim 3 [pers. comm. by H. van Essen (Dieren)] and Voigtstedt (Dietrich 1965)[19] (cf. Fig. 79), so that spatial overlap between advanced *meridionalis* populations and *M. trogontherii* cannot be excluded.

10. *Panthera onca gombaszoegensis* characterizes the Early Pleistocene post-Villafranchian as well as almost the entire early Middle Pleistocene in Europe (Stránská Skála, Süssenborn, Westbury-sub-Mendip, Mosbach 2, Petralona, Hundsheim, Artenac, l'Escale) (Hemmer & Schütt 1969, 1970; M.-F. Bonifay 1971, 250ff.; Hemmer 1971, 1972; Bishop 1982, 51-54; Tournepiche 1984, 134-135, 1993, 35ff., 1994, 27ff.; Debenath & Tournepiche 1993, 30-31; overview in Hemmer et al. 2003). In northwestern Europe, possibly post-Elsterian finds may have been found at Swanscombe (A. Turner 1999, 178-179), and in southeastern Europe at Vértesszölös 2 (Jánossy 1986, 83). The jaguar finds from Untermassfeld therefore cannot contribute to a sharper definition of the site's minimum age.

11. In an evolutionary sense, the Untermassfeld cheetah *Acinonyx pardinensis pleistocaenicus* is at the threshold of a broad transitional zone that leads up to the recent *A. jubatus* (Hemmer 2001). For the time being it remains unclear whether the cheetah from the early Middle Pleistocene of Hundsheim (with *Arvicola mosbachensis*)[20], which was described

[19] The identity of the mammoths found at Voigtstedt is a subject of current debate (summarized in Lister et al. 2005).

by Thenius (1954) as *Acinonyx intermedius*, had already further approached early *jubatus* populations (Hemmer l.c., 725).

12. The lynx remains from Untermassfeld (*Lynx issiodorensis* ssp. ex aff. *spelaeus*) are smaller than the Central-European finds from Mauer and Mosbach 2, which likewise still belong to the *issiodorensis* group of forms (Hemmer 2001). Until now there is only an isolated M^1 from Atapuerca (TD 6) (Garcia & Arsuaga 1999, 423).

13. *Megantereon cultridens adroveri* from Untermassfeld currently represents the latest known occurrence of European dirktooth cats.

14. According to Hemmer (2001) the Untermassfeld sabretooth cat *Homotherium crenatidens* (with *nestianum* morphotypes) is to be separated from the predominantly Middle Pleistocene species *H. latidens* (without *nestianum* morphotypes) (Steinheim, Montmaurin, Grotte de la Baume, Artenac) (Kurtén 1968, 77; Beden et al. 1984).

15. In Europe the hyaena *Pachycrocuta brevirostris*, abundantly documented at Untermassfeld, survived into the early Middle Pleistocene (e.g., Süssenborn, with *Mimomys savini*) (A. Turner & Antón 1996). Until now no morphological changes have been known to occur within the Late Villafranchian to early Middle Pleistocene *Pachycrocuta* populations of the Western Palaearctic.

16. The so far youngest evidence of the badger *Meles hollitzeri*, which is also known from Untermassfeld, was produced from the early Middle Pleistocene layers 6 and 4b of the Treugol'naya Cave in the northern Caucasus (Baryšnikov 1993; Wolsan 2001).

17. According to preliminary conclusions by Garcia (2004), *Ursus* cf. *dolinensis* from Untermassfeld is closely related to the late Early Pleistocene fossil populations of Le Vallonnet and Atapuerca-Gran Dolina (TD4). Within the cave bear lineage it therefore distinguishes itself from the early Middle Pleistocene *U. deningeri* (Mosbach 2, Mauer, Scharzfeld, Petralona B, Westbury-sub-Mendip).

18. The variable shape of the M_1 metaconid and some other dental features of *Canis (Xenocyon) lycaonoides* from Untermassfeld more closely correspond to those of older fossil populations (cf. Section 7.2.1.) than to those of the European (early) Middle Pleistocene (Stránská Skála, Koněprusy, Gombasek, Würzburg-Schalksberg, Mosbach 2, Vertesszölös 2) (Sotnikova 2001).

19. To judge by the length ratios of P^4 and M^{1-2} as well as by the relation between radius and humerus, the form of the wolf *Canis mosbachensis* from Untermassfeld is more primitive in appearance than the "late Mindel" (M.-F. Bonifay 1971, 350ff.) fossil population from l'Escale (Sotnikova 2001).

20. Within the small mammal association of Untermassfeld, the evolutionary levels of *Microtus* and *Mimomys savini*, the occurrence of *Mimomys pusillus*, and the evolutionary advancement of *Spermophilus* and *Glis sackdillingensis,* indicate a greater age than the following European faunas of the latest Early to the early Middle Pleistocene: Atapuerca TD 5, Slivia, Isernia la Pineta, Voigtstedt, West Runton, Kozi Grzbiet, Süssenborn, Koněprusy C 718, Gombasek (*pro parte* Maul et al. 1998; Maul 2001a).

The stratigraphical range of many mammalian taxa of the Untermassfeld fauna extends across the M/B boundary and far into the early Middle Pleistocene (cf. Fig. 79 and also Maul 2001a, fig. 54). Apart from the small mammal associations (e.g., from West Runton, Voigtstedt, Přesletice; Maul 2001a, fig. 54), the finds especially of *Bison menneri*, *Capreolus cusanoides*, *Eucladoceros giulii*, and *Stephanorhinus etruscus* (see above) allow a clear chronological separation of the Untermassfeld fossil community from European early Middle Pleistocene *Mimomys savini* faunas just beyond the M/B boundary. In western and Central Europe (West Runton, Voigtstedt, Süssenborn, Koněprusy) each of the species mentioned had already been replaced by more advanced forms (various species of *Bison*, *Capreolus suessenbornensis*, *Megaloceros savini*, *Praemegaceros*[21] *verticornis*, populations of more robust *Stephanorhinus etruscus*) during the initial phase of the Brunhes polarity zone. Furthermore, the elaphine deer *Cervus acoronatus* is regularly found above the M/B boundary, in contrast with Untermassfeld (H.-D. Kahlke 2001a, 476).

Large mammal communities fundamentally

[20] On the use of the name *Arvicola mosbachensis* (Schmidtgen, 1911) instead of *Arvicola cantianus* (Hinton, 1910) see Maul et al. (2000).

[21] See Abbazzi (2004) on the validity of the generic name.

similar to those of the beginning of the Brunhes zone may also be traced directly below the M/B boundary, from Central Europe as far as Transcaucasia at least. From Dorn-Dürkheim 3 (with *Mimomys savini*), for example, there is evidence of *Mammuthus trogontherii, Bison schoetensacki, Capreolus suessenbornensis, Alces latifrons, Praemegaceros verticornis*, and also of *"Cervus" reichenaui* and *C. acoronatus* (Franzen 1999a, 1999b; Franzen et al. 2000). The majority of the elephantid dental remains are referable to *M. meridionalis* ssp. [pers. comm. by H. van Essen (Dieren)]. The fauna from Akhalkalaki (southern Georgia) (Vekua 1962, 1986; updated faunal list in Hemmer et al. 2001) is comparable with Dorn-Dürkheim 3 with respect to chronology, and also in an ecological sense because of a distinctly expressed open ground component. The Georgian site represents an animal world that was apparently slightly further adapted to dry conditions[22]. The characteristic forms *Praemegaceros verticornis, Stephanorhinus etruscus*, and *Equus suessenbornensis* are common to both sites. More recently a new excavation level was made in the upper part of the fossiliferous horizon of Akhalkalaki (M. K. Gabunia et al. 1994), and it yielded new vertebrate finds. In this section reversed magnetization was recorded (Tappen et al. 2002), so that the above-mentioned faunal remains described by Vekua (1962, 1986) are probably likewise ascribable to a late part of the Matuyama polarity zone.

European small mammal faunas from horizons within which the M/B boundary was demonstrated (Únètice, Mahlis) may without any problem be classified as younger than Untermassfeld (Maul 2001a, 874).

The existence of a noticeable time interval between the Untermassfeld fauna and the M/B boundary is also made apparent by the research results on the mammal finds from Atapuerca (Trinchera Dolina succession, TD) (summarized by Cuenca Bescós et al. 2004). Here the M/B boundary was demonstrated within TD 7 on the basis of compelling evidence (Parés & Pérez-González 1998, 58, 1999). The bison remains from ?TD 3 and TD 6 are more evolved than *Bison menneri* from Untermassfeld (see above). On the basis of small mammals, Maul (2001a, 874) concluded that Atapuerca TD 5 is younger than Untermassfeld.

European small mammal faunas from the time span between the Jaramillo polarity subzone and the M/B boundary (Samin, Karaj Dubina) likewise do not contradict a positioning of the Untermassfeld fossil bed below the M/B boundary.

A further indication that the stratigraphical position of the Untermassfeld fauna lies definitely below the M/B boundary may be provided by the remains of *Hippopotamus amphibius antiquus*. Compared to the skull find from Untermassfeld (R.-D. Kahlke 2001e), the *Hippopotamus* skull from S. Oreste, which according to Mazza (1991, 168) should be positioned between Jaramillo and the M/B boundary, turns out to be more advanced morphologically.

Since larger find complexes from the inversely polarized time span between Jaramillo and the M/B boundary are lacking, especially in Central Europe, the possibilities for the Untermassfeld large mammal fauna to produce evidence of the minimum age of the site are still limited. Information may arise from a comparison with new finds from Trinchera Dolina and the deeper part of the Sima del Elefante of Atapuerca (Rosas et al. 2004); Barranco Léon 5 (Martinez et al. 2003); Incarcal (Galobart & Maroto 2003); the deeper horizons of Stránská Skála (see Valoch 1995, 73 on the polarity); Akhalkalaki in southern Georgia (Vekua 1962, 1986; R.-D. Kahlke 1985b; Hemmer et al. 2001; Tappen et al. 2002); and the southern Tadzhikian fauna from Lakhuti 2 [Nikiforova & Vangengejm (eds.) 1988; Sotnikova 1989; Sotnikova et al. 1997, 253-254] (cf. Section 7.4.).

The age of the M/B boundary was stated to be 775 ka $\pm$ 10 ka BP (Bassinot et al. 1994) or 780 ka BP (Berggren et al. 1995, table 7; Shackleton 1995, table 17.1.). Based on these data, combined U-series / ESR dating resulted in absolute age values between 780 and 886 ka BP for Atapuerca TD 6 (below the M/B boundary) (Falguères et al. 1999, fig. 1). Allowing for the age difference between Atapuerca TD 6 and TD 5, and additionally for the interval that according to palaeontological evidence separates it from Untermassfeld (see above), a rough estimate of 0.9 Ma may be given for the minimum age level of our site.

7.3. Stratigraphic position and correlation

The time span of < 1.2 - 0.9 Ma BP, determined for the deposition of the fossiliferous sands of Untermassfeld on the basis of the faunal comparisons made so far (cf. Sections 7.2.1., 7.2.2.) allows a more precise statement regarding the litho-magnetostratigraphical position-

[22] *Hippopotamus* represents an African immigrant. It had probably migrated via the Levantine corridor (Vekua 1959; R.-D. Kahlke 1985). It was strongly tied to streams and adjacent meadow areas (R.-D. Kahlke 1997b, 365ff.).

ing of the site (cf. Section 7.1.). It seems fairly unlikely that the change from reversed to normal polarity observed in the Lower Fluviatile Sands of Untermassfeld should be correlated with the M/B boundary. On the contrary, as far as known today the magnetic boundary that lies directly underneath the bones probably corresponds to the base of the Early Pleistocene Jaramillo polarity subzone. According to Berggren et al. (1995, table 7) and Shackleton (1995, 243), the Jaramillo begins at 1.07 Ma BP. Hence an absolute age of slightly over one million years is to be attributed to the faunal remains from Untermassfeld.

Shackleton (1995, fig. 17.1.) placed the warm fluctuations of marine oxygen isotope stages (MIS) 27-31 within the Jaramillo. If one takes the lithological and magnetic characteristics of the bone-bearing Lower Fluviatile Sands of Untermassfeld into consideration, a correlation with MIS 31 – a pronounced warm interval that overlaps the base of the Jaramillo – would be the likeliest assessment of the time at which the fossil lagerstätte was formed. Gibbard et al. (1998, 246) correlated MIS 31 with the earlier Bavelian.

In terms of the finer subdivision of the Bavelian produced by Zagwijn & Jong 1984 (166-167, fig. 9), the formation of the fossil lagerstätte of Untermassfeld would have taken place during the normally magnetized part of the Bavel (s. str.) Interglacial (Bv 2-4, with successive immigration of thermophilous tree vegetation into the area of the Netherlands). This time span may be correlated with the early Galerian of the Italian Mammal Ages (Abbazzi et al. 1995, fig. 3; Sardella et al. 1998, fig. 2), or with a late "Protogalerian" *sensu* Caloi & Palombo (1995, fig. 1). Within the biozonal subdivision of Guérin (1990), Untermassfeld is to be placed in an early section of MNQ 20. After Agusti et al. (1987), the fauna corresponds with the zone M_mQ-3a. However, the definitions of both biozones do not entirely do justice to the level of development observed in the composition of the Untermassfeld fauna. After the Russian subdivision (Vangengejm et al. 1991), the fauna from Untermassfeld would have to be attributed to the earlier Taman (= middle Apsheronian of the Ponto-Caspian area). On the basis of the small mammal fauna, Untermassfeld belongs to the late Early Biharian sensu Fejfar & Heinrich (1981) (Maul 2001a).

7.4. Supraregional comparisons

To illustrate the stratigraphical position of the Untermassfeld fauna, important large mammal associations from Europe to Middle Asia and the Near East, are presented below in their age relation to Untermassfeld[23]. In most cases the time span from Late Villafranchian to early Middle Pleistocene is considered, so as to chronologically flank the formation of the Untermassfeld fossil locality.

In Spain the Plio- / Pleistocene boundary is marked by the site **Fonelas P-1**, which was discovered just a few years ago (Arribas et al. 2001). *Gazellospira* aff. *torticornis*, *Leptobos etruscus*, *Mitilanotherium* aff. *martinii*, *Sus strozzii*, *Canis etruscus*, and other species (Arribas et al. 2004) demonstrate the unmistakably greater age of this fossil assemblage in comparison with Untermassfeld. Finds of Pleistocene faunas below the Jaramillo polarity subzone are also known from the **Cueva Victoria** [Agusti 1991; Agusti & Moyà-Solà 1991; Campillo et al. (eds.) 1992; Cerdeño 1993]. Among the large mammals, *Equus stenonis stenonis*, *Mammuthus meridionalis*, and *Canis etruscus*, for example, suggest that this site is older than Untermassfeld. With, among other taxa, *Mammuthus meridionalis* and *Canis etruscus*, the rich fossiliferous deposit of **Venta Micena** [Moyà-Solà et al. (eds.) 1987; Agusti & Moyà-Solà 1991, 1992; Campillo et al. (eds.) 1992; Cerdeño 1993; Martinez Navarro & Palmqvist 1995; Palmqvist et al. 1996, 1999; Palmqvist 1997; Torres et al. 1997; Arribas & Palmqvist 1998; Martinez-Navarro et al. 1998; Crégut-Bonnoure 1999; Eisenmann 1999] yielded faunal elements that indicate a greater age than Untermassfeld. The site **Fuente Nueva 3**, with, e.g., *Mammuthus meridionalis*, is likewise to be positioned below the Jaramillo (Martinez-Navarro et al. 1997; Crégut-Bonnoure 1999).

The entire TD (Trinchera Dolina) succession of **Atapuerca-Gran Dolina** as it has so far been treated [Carbonell et al. 1995; Bermúdez de Castro et al. 1997; Aguirre (ed.) 1998; Made 1998, 1999, 2001; Falguères et al. 1999; Garcia & Arsuaga 1999, 2001a; Garcia Garcia 2003, Cuenca Bescós et al. 2004; also J. Human Evol. 37, 3/4 on hominid remains] is younger than the fossiliferous deposit of Untermassfeld (faunistic indications are discussed in Section 7.2.2.). The normal polarization that has been observed once within TD 1 (Parés & Pérez-González 1999, 332) is with reservation related to the Jaramillo. The fauna from **Huescar 1** (Mazo et al. 1985; Cerdeño 1993), which together with Atapuerca TD 4-6 was assigned to the zone M_mQ-3b (Made 1999, table 1), might likewise be younger than Untermassfeld. However, results supporting this are controversial [oral comm. by N. Garcia (Madrid), March 2005].

[23] Similar comparisons involving relevant small mammal faunas are to be found in Maul (2001a, 866ff.).

The fauna from **Cúllar de Baza 1**, with, among other species, *Cervus elaphus*, *Megaloceros savini*, *Equus suessenbornensis*, *E. altidens*, and *Crocuta crocuta*, is to be positioned above the M/B boundary (Cerdeño 1993; Agusti & Oms 1998; Alberdi & Sánchez-Chillón 1998; Agusti et al. 1999).

From F r a n c e also, several sites are known that offer a chronological framework for the stratigraphical positioning of Untermassfeld. The final phase of the Late Villafranchian is characterized by the fauna from **Sainzelles**, which includes, for example, *Equus stenonis* and *Mammuthus meridionalis* (M.-F. Bonifay 1971, 349, 1988, 184, 1990, 477, 1992, 58; Geraads 1990, 170). Its absolute age was stated to be approximately 1.4 Ma BP (M.-F. Bonifay & Brugal 1999, 558). **Ceyssaguet**, which contains, e.g., *Equus stenonis*, *Ursus etruscus*, and *Canis etruscus* (M.-F. Bonifay et al. 1984; M.-F. Bonifay 1986, 274ff., 1988, 184, 1993, 113; Geraads 1990, 169-170; Aouadi & M.-F. Bonifay 1997-1998), is likewise older than the finds from Untermassfeld. Meanwhile, absolute age indications lie around 1.2 Ma BP (M.-F. Bonifay & Brugal 1999, 558). It seems probable that the large mammal remains from **Courterolles** (Brochet et al. 1983), which up to now has been cautiously correlated with the Waalian, are to be chronologically bracketed together with Ceyssaguet.

The large mammal remains from the cave of **Le Vallonnet** (Lumley et al. 1988; Moullé 1990, 1997-1998, 1999; Hemmer et al. 2004) originate from normally-polarized horizons attributed to the Jaramillo. Composition and stage of development of the local fauna more or less correspond to those found at Untermassfeld (Moullé et al. 2000). A revision of the fauna found at **Saint-Prest** (Guérin et al. 2003) likewise showed up certain similarities with Untermassfeld. Within MNQ zone 20 (cf. Section 7.3.), *Bison schoetensacki* and possibly an elaphine deer could demonstrate a slightly younger age in comparison with Untermassfeld. Nevertheless, the large deer of this fauna is a *Eucladoceros*, which according to A. Lister (pers. comm., February 2006) could well be conspecific with *E. giulii* from Untermassfeld.

The faunas from **Soleilhac** (M.-F. Bonifay 1986, 270ff., 1988, 184, 1993, 114; E. Bonifay & Mergoil 1988; Geraads 1990, 170-171) were for a long time also assigned to the Jaramillo. However, the excavation area Soleilhac-Centre, in particular, yielded a fauna which is demonstrably younger than that of Untermassfeld because it includes, e.g., *Bison schoetensacki* and *Capreolus suessenbornensis*. More recent datings of the site have suggested 500-600 ka BP (Raynal et al. 1995, 141). To judge by the evolutionary level of a large bovid close to *Bison schoetensacki*, the faunal remains found at **Durfort** are likewise younger than Untermassfeld. With some reservation, Brugal (1995) chronologically placed them between Le Vallonnet and Isernia la Pineta (see below).

Within France, several rather large early Middle Pleistocene faunas are known from above the M/B boundary, for instance that from **Vergranne**, which comprises, among other species, *Capreolus* cf. *suessenbornensis*, *Equus mosbachensis*, *Ursus deningeri*, and *Canis mosbachensis* (Campy 1983), from **l'Escale** (Saint-Estève-Janson), likewise with *C. suessenbornensis*, *E. mosbachensis*, and *C. mosbachensis* (M.-F. Bonifay 1971, 1981; Crégut-Bonnoure 1987; Alberdi et al. 1997, 127), and from **Lunel Viel** (Grotte du Mas des Caves) (M.-F. Bonifay 1971; Brugal 1985; Eisenmann et al. 1985; Fosse 1996; Alberdi et al. 1997, 127), which shows a similar range of species indicating an appreciably higher stratigraphical position when contrasted with Untermassfeld. The younger age of the succession at **Tautavel** (La Caune de l'Arago) is demonstrated by, e.g., *Praeovibos priscus*, *Rangifer tarandus* ssp., *Cervus acoronatus*, and *Equus mosbachensis* (Crégut & Guérin 1979; Crégut 1980a, 1980b, 1981; Moigne 1984; Eisenmann et al. 1985; Alberdi et al. 1997, 127; Moigne et al. 2000). Something similar goes for the succession within the lower profile of **Artenac** (Beden et al. 1984; Tournepiche 1984, 1985, 134-136, 1993, 35ff., 1994, 27ff.; Delagnes et al. 1999).

Typical evidence of the Early Pleistocene animal world of I t a l y was produced by the Late Villafranchian Faunal Units of Tasso and Farneta (Azzaroli 1977) (surveys in Torre et al. 1993; Sardella et al. 1998). The Tasso complex comprises the finds from **Casa Frata**, with, e.g., *Leptobos*, *Eucladoceros*, *Mammuthus meridionalis*, *Ursus etruscus*, and *Canis etruscus* (Giuli & Masini 1987), as well as the sites **Villa S. Faustino** and **Colle S. Andrea** (Ambrosetti et al. 1995; Sardella et al. 1995) in the Tiberine basin. Chronologically close are also the faunal remains from the karstic fissure of **Alonte** near Vicenza (Masini et al. 1995). Here *Leptobos* cf. *etruscus* and *Ursus* cf. *etruscus* likewise indicate an age that is markedly greater than that of Untermassfeld. Evidence of both Tasso and Farneta faunas is known from the **Mugello basin** (Abbazzi et al. 1995). The local faunas of **Pietrafitta** (Azzaroli & Mazza 1993; Mazza et al. 1993; Ferretti 1999, 507; summary in Masini et al. 1994, 385; Gentili et al. 1996, 188) and **Selvella** (Giuli 1987) – with *Leptobos* demonstrated in each of them – belong to the Farneta Faunal Unit. The fauna from **Pirro Nord** (= Cava Pirro) (Giuli et al. 1987; Masini et al. 1994, 385), possibly to be correlated with the Menapian, follows with an *Eobison* and with cervids that are older in comparison with those of Untermassfeld (cf. Section 7.2.1.).

The local fauna of **Colle Curti**, which in contrast with Untermassfeld already contains *Praemegaceros verticornis* as a more modern element, was placed within the Jaramillo polarity subzone (Masini et al. 1994, 386; Caloi & Palombo 1995, 394ff.; Torre et al. 1996, 565ff., 1999, 541f.; Ferretti 1999, 508).

The time span between Jaramillo and the M/B boundary is represented by the fauna from **Imola**, with *Bison schoetensacki*, *Praemegaceros verticornis*, and *Mammuthus meridionalis* (Masini et al. 1995; Ferretti 1999). Chronologically close to the M/B boundary is the **Slivia** Faunal Unit with *Mimomys savini* (Ambrosetti et al. 1979; Sardella et al. 1998, 502; Torre et al. 1999, 542). The Isernia Faunal Unit follows in the lower part of the Brunhes polarity zone and comprises the sites **Isernia La Pineta** (Cremaschi et al. 1985; Sala 1987, 1990; Mussi 1995, 29f.; Gagnepain et al. 1996, 130; Koenigswald & Kolfschoten 1996, 216-217) – which beside *Arvicola mosbachensis* contains, e.g., *Bison schoetensacki*, *Panthera leo fossilis*, and *Ursus deningeri* – and **Ponte Galeria** (Caloi & Palombo 1980, 1986, 107) with *Megaloceros savini*, *Praemegaceros verticornis*, and *Elephas (Palaeoloxodon) antiquus*. The fauna from **Venosa**, which includes *Equus altidens*, is still younger in the early Middle Pleistocene (Alberdi et al. 1988, 34ff.; Mussi 1995, 31; Sardella et al. 1998, 502), as does **Fontana Ranuccio** with *Bos primigenius, Dama clactoniana, Equus* cf. *mosbachensis*, and *Elephas (Palaeoloxodon) antiquus* (Caloi & Palombo 1986, 108f.; Mussi 1995, 33).

In Great Britain, large mammal remains from the earliest Early Pleistocene were recovered from the Weybourne Crag (Pre-Pastonian, Pastonian), for example near **East Runton** (Stuart 1974, 234f., 1988, 57f.; Gibbard et al. 1991, 29ff.), which is to some extent comparable to the Tegelen Clays (see below) (Lister 1998). Among other taxa, *Alces gallicus*, *Eucladoceros tegulensis* (= *senezensis* = *ctenoides*[24]), *Equus stenonis*, and *Mammuthus meridionalis* here indicate the older character of the fauna with respect to Untermassfeld.

Better known are the many early Middle Pleistocene, and hence in comparison with Untermassfeld younger, faunas of England. On the basis of the occurrence of *Mimomys savini*, an older group comprising the sites **West Runton** (West Runton Freshwater Bed as part of the Cromer Forest-bed Formation, type deposit of the Cromerian *s.s.* Stage) (Stuart 1981, 1992, 1996, 11ff., 1997; Lister 1993, 1996b, 27ff.; Ashwin & Stuart 1996), **Sidestrand** (Stuart 1996, 14), **Little Oakley** (Lister et al. 1990; Stuart 1996, 16ff.), and **Sugworth** (Stuart 1980, 1988, 1996, 18) may be distinguished. When compared with their counterparts found at Untermassfeld, many elements of the West Runton fauna demonstrate their distinctly younger stratigraphical position, e.g., *Bison schoetensacki*, *Alces latifrons*, *Cervus elaphus*, *Megaloceros savini*, *Praemegaceros verticornis*, a caballoid equid, and *Crocuta crocuta*. Something similar goes for the faunal ranges of the other localities mentioned, which are less abundant in species. Younger, pre-Anglian large mammal faunas with *Arvicola mosbachensis* were recovered from **Westbury-sub-Mendip** (Bishop 1982; Roberts et al. 1995, 173f.; Stuart 1996, 19; A. Turner 1999), **Boxgrove** (Bowen & Sykes 1994; Roberts et al. 1994, 1995, 171f.; Stuart 1996, 19; Schreve et al. 1999), and **Ostend** (Koenigswald & Kolfschoten 1996, 217; Stuart 1996, 14ff.).

Among the large quantities of vertebrate remains that were recovered from the bottom of the North Sea between the British Isles and the Netherlands (e.g., the Deep Water Channel), a late Early to early Middle Pleistocene component (**Faunal Association 2**) is found which is predominantly referable to the later Bavelian and the Cromerian Complex (summarized by Kolfschoten & Laban 1995, 143-146; Vos et al. 1999; Kolfschoten & Essen 2004). Some finds (e.g., of *Hippopotamus amphibius antiquus* and a large bovid) might approach or correspond to the respective stages of development observed at Untermassfeld (cf. also Mol 1998, 43; Drees 1999).

In the Netherlands the late Middle Villafranchian fauna from the **Tegelen** Clays, with, among other taxa, *Leptobos*, *Eucladoceros tegulensis*, *Sus strozzii*, and *Pliocrocuta perrieri* (summary in Vervoort-Kerkhoff & Kolfschoten 1987; Kolfschoten 2001; bibliography in Meijer 1996), obviously distinguishes itself from the Untermassfeld association.

More relevant are the large mammal associations from the **Maasvlakte** – an artificial terrain raised with a sand-in-water slurry downriver from Rotterdam – and Dorst-Surae. Within the older component of the mammalian remains recovered out of context from the Maasvlakte, Kolfschoten & Vervoort-Kerkhoff (1999a, 1999b) distinguish a

[24] See Vos et al. (1995, 109ff.) and H.-D. Kahlke (1997, 226) on the synonymy.

Fauna 0, which is correlated with the Eburonian / Waalian. A find of *Soergelia* probably belongs to this group (Kolfschoten & Vervoort-Kerkhoff 1999b, 377ff.). The material subsumed under **Maasvlakte Fauna 1** is chronologically closest to Untermassfeld. It is comparable to the Faunal Association 2 of the North Sea bottom and is referred to the time span from the Bavelian to the Cromerian Complex by Kolfschoten & Vervoort-Kerkhoff (1999a, 1999b). It would be interesting to compare the cervids, hippopotamids, suids, rhinocerotids, equids, and felids known from the Maasvlakte faunas 1-0 (on this subject: Hemmer & R.-D. Kahlke 2005) with corresponding finds from Untermassfeld. The faunule from **Dorst-Surae**, which includes an evolved form of *Mammuthus meridionalis* and *Eucladoceros*, belongs to a later section of the Bavelian Complex (the Leerdam Interglacial) (Kolfschoten 1990a, 267-269; Mol et al. 1999, 126-127; Essen 2003, *passim*).

Evidence of Pleistocene large mammal faunas situated below the Jaramillo polarity subzone is hardly known from G e r m a n y. According to Koenigswald & Tobien (1987), the finds from **Mosbach 1** [= "Grobes (Coarse) Mosbach"] are to be placed here (elaborate researches into the history of finds and the relevant literature are to be found in H.-D. Kahlke 1961). However, some of the taxa subsumed under the Coarse Mosbach (Koenigswald & Tobien 1987, 230) seem to contradict the age assessment deduced from the results on palaeomagnetism (e.g., *Praemegaceros verticornis*, elaphines).

On the basis of both the stratigraphical context of the entire profile, and the interpretation of a palaeomagnetical fluctuation as belonging to the Jaramillo subzone, **Kärlich A** – with *Hippopotamus* – is probably to be chronologically correlated with Untermassfeld (Kolfschoten & E. Turner 1996, 232f., 246).

At **Dorn-Dürkheim 3** a *Mimomys savini* fauna is known to occur below the M/B boundary (Franzen 1999a, 1999b; Franzen et al. 2000). Its large mammal assemblage approaches those of the earliest Middle Pleistocene (Voigtstedt, Süssenborn) and hence characterizes it as distinctly younger with respect to Untermassfeld (cf. Section 7.2.2.). At **Würzburg-Schalksberg** (Schütt 1974; Mäuser 1987; R.-D. Kahlke 1989), data on palaeomagnetism and finds of arvicolids are absent. Here, too, the joint occurrence of *verticornis* deer and a cervid close to *Eucladoceros giulii* from Untermassfeld (Mäuser 1987, 27-31; H.-D. Kahlke 1997, 225-227) leads one to suspect a stratigraphical position below the M/B boundary (cf. also Atapuerca TD4 and TD6 mentioned above).

From the early Brunhes polarity zone, very extensive early Middle Pleistocene faunas including an evolved *Mimomys savini* are known from **Voigtstedt** [H.-D. Kahlke (ed.) 1965] and **Süssenborn** [H.-D. Kahlke (ed.) 1969; Forstén 1986]. The succession of the *Mimomys savini* faunas from **Kärlich C**, **E**, and **F**, which include sporadic large mammal finds, is to be positioned above the M/B boundary (Koenigswald & Kolfschoten 1996, 218; Kolfschoten & E. Turner 1996, 232ff.; Koenigswald & Heinrich 1999, 77-78).

Several mammalian faunas were shown to belong to a yet later part of the early Middle Pleistocene through the presence of *Arvicola mosbachensis*. To these belong **Kärlich G** (Kolfschoten 1990b, 28; Kolfschoten & E. Turner 1996, 231ff.) and **Miesenheim 1** (E. Turner 1990; Kolfschoten & E. Turner 1996, 232ff.). In each case the absolute age has been estimated at > 453 ± 7 ka BP on the basis of a dated pumice layer within the overlying deposits (Bosinski 1995, 106). Further large mammal faunas of the pre-Elsterian, in part associated with *Arvicola mosbachensis*, are represented by **Mauer** (summary in Koenigswald 1992, 1997), **Mosbach 2** [= "Graues (Grey) Mosbach" with main fauna] (H.-D. Kahlke 1961; Brüning 1970, 1978; Koenigswald & Tobien 1987; Pfeiffer 1997, 1998; Hemmer et al. 2003), **Randersacker** (Rutte 1958, 1959; Nobis 1981; Guenther 1989), and **Jockgrim** (Kuss 1955, 1957; Guenther & Mai 1977), and also by the fissure infills **Erpfingen 1** and **3** (Heller 1936, 1958; Kretzoi 1965, 627-629). **Bad Frankenhausen** (H.-D. Kahlke 1963, 952, 1964; Fischer 1985, 81-83) represents the earliest typical Middle Pleistocene cold-stage fauna with *Mammuthus trogontherii* and *Coelodonta antiquitatis* ssp. (R.-D. Kahlke 1994, 86), which may be positioned even before the maximal advance of the Elsterian ice.

In A u s t r i a the cave and fissure system of **Deutsch-Altenburg** is of special importance. Here the sites **DA 2C$_1$** and **DA 4B** yielded small and large mammal assemblages abundant in species and of an Early Pleistocene age greater than that of Untermassfeld (Rabeder 1976; Mais & Rabeder 1984; Fladerer 1987; Frank & Rabeder 1997a; Pfeiffer 2000).

Younger than Untermassfeld is the early Middle Pleistocene *Arvicola mosbachensis* fauna from **Hundsheim**, which includes *Bison schoetensacki*, *Cervus elaphus*, *Equus mosbachensis*, *Panthera pardus*, and *Crocuta* sp. (Daxner 1968; Rabeder 1972; Crégut-Bonnoure 1987; Frank & Rabeder 1997b).

In C z e c h i a the Early Biharian fissure or

cave infill of **Chlum 6** near Srbsko (Bohemia), which also yielded a series of large mammal taxa (Fejfar 1961, 112ff.; Fejfar & Horáček 1983, 127-128), turns out to be somewhat older than the Untermassfeld fauna, as evidenced by the arvicolids (Maul 2001a, 870).

A very complex site with a long research history is the **Stránská Skála** mountain near Brno (Moravia), which consists of karstified Jurassic limestones. There are older collections from quarried karstic cavities and a talus cone, all containing a rich quantity of early Middle Pleistocene faunal components [Musil (ed.) 1971, with many separate discourses]. More recently, the collections made from horizontal sections of the talus cone profile of the Stránská Skála have assumed importance [Musil (ed.) 1995a]. The Jaramillo polarity subzone proved demonstrable within the lower section of this profile (layers 4, 4a). The M/B boundary lies within the upper part of layer 5 and the lower part of layer 6 (Valoch 1995, 73). The following large mammals from the **talus cone of the Stránská Skála** above the M/B boundary (Musil 1995b; H.-D. Kahlke 1995b) confirm that each of these levels is younger than Untermassfeld: *Bos primigenius* (layer 14), *Capreolus* cf. *suessenbornensis* (layers 8-10), "Cervus" cf. *reichenaui* (layer 9), *Cervus* sp. (*acoronatus* / *elaphus* group) (layers 6-15), *Praemegaceros* sp. (layers 10b, 13), *Crocuta crocuta* (layer 13), *Ursus deningeri* (layer 11-14a).

The infill of the cave C 718 near Zlaty Kun (Bohemia), usually called **Koněprusy C 718**, contained a multifarious early Middle Pleistocene vertebrate fauna with *Mimomys savini* (Late Biharian) (Fejfar 1961, 113ff.; H.-D. Kahlke 1975, 333-334). Among the large mammals particularly *Bison schoetensacki*, *Cervus acoronatus*, *Panthera pardus*, *Pliocrocuta perrieri*[25], and *Ursus deningeri* are indicative of its age, which is markedly younger than Untermassfeld. The latest Biharian, with very advanced forms of *Mimomys savini*, is known from a fluvio-limnic residual part of the Elbe river system at **Přezletice** near Prague (Fejfar 1969, 1993; Koenigswald & Kolfschoten 1996, 214f.). The chronological distance from Untermassfeld is here documented by, e.g., *Capreolus* cf. *suessenbornensis*, *Cervus* cf. *elaphus*, and *Equus mosbachensis*.

Holec (1985) gave a survey of Quaternary sites in Slovakia. To judge by its small mammal assemblage, the fissure infill **Včeláre** 3 in the southeast of the country is older than Untermassfeld. There is evidence of some large mammals there as well (Horáček 1980; Fejfar & Horáček 1983, 118-120). The profile extends from its beginnings in the Villanyian into the Early Biharian (Eburonian) (Horáček 1985, 356).

The cave ruin of the site **Gombasek 1** (= Gombaszög) in the Southern Slovakian karst has become known particularly because of its carnivore association, which is rich in species (Kretzoi 1937/38, 1941a). With respect to Untermassfeld, a younger age of the late Biharian fauna (Fejfar in Bartolomei et al. 1975, 411; Maul 2001a, 870) is here indicated by, e.g., *Ursus deningeri*, but also by the ungulate fauna with *Bison schoetensacki*, *Capreolus suessenbornensis*, and *Alces latifrons* (Fejfar 1961, 116; Kretzoi 1965, 616f.). A similar chronological position may have to be assigned to the karstic infill of **Žirany** (summary in Holec 1985, 121f.).

An inventory of the Plio- / Pleistocene faunas of Poland is to be found in the survey edited by Kowalski (1989). The Early Biharian fauna from the **Żabia Cave** near Kroczyce yielded a series of remains of smaller carnivores (Wolsan 1990). These do not, however, allow a direct comparison with the Untermassfeld material. To judge by its small mammals (Maul 2001, 868f.), the Żabia association is older than that from our site. The presence of *Alces gallicus* in the Żabia Cave as suggested by Czyzewska (1989, 211, 213) tallies with this idea.

Because of the presence of *Capreolus suessenbornensis*, *Alces latifrons*, and *Praemegaceros verticornis* (Czyzewska 1989, 211, 214), as well as of *Crocuta* sp. and *Ursus deningeri* (Wolsan 1990), the large mammal component of the Late Biharian fauna from **Kozi Grzbiet** near Kielce shows that its age is younger than that of Untermassfeld.

The Quaternary sites of Hungary were summarily described by Jánossy (1986). The many very abundant and well-researched small mammal occurrences in the country are not always matched by large mammal finds.

Markedly older than Untermassfeld are the multifarious faunas of the Late Villanyian (late Middle Villafranchian) from sites such as **Villány 3** (updated fauna list in Jánossy 1986, 32-34) with, for example, *Gazellospira* cf. *torticornis*; **Osztramos 3** (Jánossy 1970, 1986, 35-36) with a representative of the *Sus strozzii* group; and **Villány 5** (Kretzoi 1956, 187ff.; Jánossy 1986, 37-38; Maul 2001a, 870) and **Kisláng** (Kretzoi 1954; Jánossy 1986, 43-45) with, e.g., *Leptobos*

[25] On the stratigraphical distribution of *Pliocrocuta perrieri* as well as on its displacement from the European faunal system during the time span between 1.5 and 0.7 Ma BP, see A. Turner (1992b, 155ff.).

etruscus and mastodonts. According to its small mammal fauna, the fissure infill **Osztramos 8**, which includes a large *Eucladoceros* (Jánossy 1972, 1986, 51-52), is likewise somewhat older than Untermassfeld (Maul 2001a, 870f.).

To judge by the rodents (Maul l.c., 871), the age of the Early Biharian cavern infill **Somssich-hegy 2** (Jánossy 1983, 1986, 54-57) with *Mimomys pusillus* corresponds to that of the fossiliferous deposit at Untermassfeld, or is slightly younger. The occurrence of *Capreolus* cf. *suessenbornensis* may speak for the latter interpretation.

The Late Biharian fauna from **Nagyharsány-hegy 4** (without *Mimomys pusillus*) (Kretzoi 1956, 55-57; Jánossy 1986, 53-54) may be considered younger than Untermassfeld (Maul l.c., 871). In this respect the occurrence of the Villafranchian bovids *Gazellospira* aff. *torticornis* and *Procamptoceras* cf. *brivatense* is surprising. Of a likewise Late Biharian (early Middle Pleistocene) age is the karstic infill of **Kövesvárad** in the Bükk Mountains (Jánossy 1963, 1986, 62-65). Among the large mammals, *Capreolus suessenbornensis* and *Ursus deningeri* suggest that its age is younger than that of Untermassfeld. In the travertine area near Budapest a probably likewise early Middle Pleistocene fauna with *Mimomys savini* was discovered at **Ürömhegy**. Among other taxa it contained *Hippopotamus amphibius antiquus* (Jánossy 1962, 1986, 65-66). **Budakalász** yielded a similar large mammal fauna with *Hippopotamus* (Jánossy 1962, 1986, 66-67). An elaphine deer (*Cervus* cf. *acoronatus*) here indicates a stratigraphical position higher than that of Untermassfeld.

A rich small mammal succession with *Arvicola mosbachensis* was described from the rock recess of **Tarkö** (Bükk plateau) (current fauna lists in Jánossy 1986, 74-77). The sequence Tarkö 15-1 documents the Hungarian Tarköian [= "Mindel II (III)", Jánossy 1986, 173]. Large mammals which are more modern than those at Untermassfeld, e.g., *Capreolus suessenbornensis*, *Cervus* cf. *acoronatus*, and *Panthera pardus sickenbergi*, were mainly recovered from the horizons Tarkö 4-2. The well-known sites **Vértesszölös 1** and **2** are likewise to be placed above the M/B boundary (Latham & Schwarcz 1990). The large mammal fauna from locality 2 with, for instance, *Bison priscus*, *Capreolus suessenbornensis*, *Cervus acoronatus*, *Equus mosbachensis*, *Panthera leo fossilis*, *Crocuta* sp., *Ursus deningeri*, and *Vulpes* aff. *vulpes* (Jánossy 1986, 80-83, 1990), represents a clearly younger association than the one from Untermassfeld. Something similar applies to the succession – in part possibly somewhat older – of Vértesszölös 1 (occupation site), which mostly contained rather strongly fragmented remains of, e.g., *Bison schoetensacki*, *B. priscus*, *Capreolus suessenbornensis*, *Rangifer tarandus stadelmanni* (above the cultural horizon), *Cervus elaphus* ssp., *Equus mosbachensis*, and *Ursus deningeri* (Kretzoi 1990).

Rakovec (1954) referred dental remains of *Alces* aff. *gallicus* from the **Vic** terrace near Ljubljana in Slovenia to the Late Villafranchian. A Middle Pleistocene age was attributed to the find of a black bear *(Ursus thibetanus)* from breccias of **Vrhovlje** (Rakovec 1975, 262).

In the site **Šandalja 1**, Croatia possesses a cave infill which is largely of Late Villafranchian age and abundant in species, containing, e.g., *Leptobos stenometopon*, *Eucladoceros tegulensis* (= *senezensis* = *ctenoides*), *Sus strozzii*, *Pliocrocuta perrieri*, *Chasmaporthetes lunensis*, *Ursus etruscus*, and *Canis etruscus* (Malez 1968, 1975, 1986, 102; Malez et al. 1992, 371-374)**,** so that it is clearly older than Untermassfeld. On the island of **Iz** (Zadar Archipelago) a series of bone breccias of apparently various ages were investigated. One of the faunal components, which includes *Leptobos* cf. *stenometopon*, *Eucladoceros*, and *Acinonyx* cf. *pardinensis*, may have to be placed close to Šandalja 1 chronologically (Malez 1969, 21/1, 1986, 103; Malez et al. 1992, 379-380).

An apparently early Middle Pleistocene age has to be assigned to a fauna from the west coast of the **Marjan** Peninsula near Split, as it comprises two ursid forms and *Capreolus suessenbornensis* (Malez 1959, 1960, 1961, 1969, 21/2). The bone breccia of **Dubci** in Central Dalmatia yielded a similar fauna (Malez 1965, 1967, 1969, 21/2), which is shown to be younger when contrasted with Untermassfeld, for instance by the occurrence of *Cervus elaphus*. After Malez (1968, 1986, 103), the lake marl of **Kninsko Polje** are attributed to the Mindel cold stage. They yielded remains of *Mammuthus trogontherii* accompanied by finds labelled as straight-tusked elephants. These obviously qualify for comparison with the small series of M_3s from Süssenborn in Thuringia, formerly identified as straight-tusked elephant but now regarded as *M. trogontherii.*

In Rumania, too, a considerable number of relatively old faunas have been recovered. In the Olteț valley of the Dacian basin the sands of **Valea Grăunceanului** (= Valea Tetoi) yielded a rich late Middle Villafranchian – and hence with regard to Untermassfeld distinctly older – mammalian fauna, which includes *Eucladoceros* sp., *Mammuthus meridionalis*, *Pliocrocuta perrieri*, and *Ursus etruscus*, as well as the giraffid *Mitilanotherium inexspectatum* (Bolomey 1965; Samson & Radulesco 1966; Samson 1976;

Radulesco & Samson 1990, 227, 1991; Radulesco et al. 1998; Spassov 2003). Likewise older than Untermassfeld are the finds from **Fîntîna Alortiței**, with *Eucladoceros* sp., *Equus stenonis mitilanensis*, a representative of the typical form of *Mammuthus meridionalis*, and *Canis etruscus*; and also the similar fauna of the **Valea Mijlociei**. An approximately Eburonian age should be assigned to these sites, as well as to **Fîntîna lui Mitilan** (Irimesti) with *Leptobos* sp., *Cervus* s.l. *nestii*, *Eucladoceros* sp., *Mitilanotherium inexspectatum*, and *Sus strozzii* (Radulesco & Samson 1962, 1985, 1991; Samson & Radulesco 1965, 1973, 1990, 227-228; Samson 1976).

The karstic infills and cave ruins of Betfia (= Püspökfürdö = Episcopia Bihorului) are of great importance. An Eburonian age was also given for the breccias of **Betfia XIII** (Terzea 1996). Here *Equus stenonis senezensis*, *Ursus etruscus*, and *Canis* cf. *falconeri* (Terzea & Jurcsák 1976) substantiate a greater age than Untermassfeld. To judge by the evidence of, particularly, its Early Biharian small mammal fauna, the classic assemblage of **Betfia II** (Kretzoi 1941b, 1965, 617-619) is likewise older than Untermassfeld (Maul 2001a, 871). The **Betfia VII** complex of strata covers the time interval from the Waalian into the Cromerian Complex (Terzea 1996). Large mammal remains were recovered from the deeper section of the profile (B-VII/1b-e). Among these are *Megantereon* and *Ursus etruscus* (Terzea 1993, 109, 1996). Considering the overall succession, the finds are older when contrasted with Untermassfeld and are placed in the later Waalian (Terzea 1995, 112). Results concerning the small mammals (*Microtus* morphotypes) suggest that the rich fauna of **Betfia V** (Kretzoi 1941b, 1965, 619-620; Terzea & Jurcsák 1968, 383-386, 1969, 206-209; Terzea 1996, table III) is probably likewise older than that from Untermassfeld (Maul 2001a, 871). Terzea (1995, 1996) placed it in the later Menapian. However, various large mammal taxa recorded from Betfia V (e.g., *Bison schoetensacki*) could indicate a younger age. Proper comparisons with the find series from Untermassfeld may shed some light here.

A group of younger fossil sites in the Olteț Valley (Dacian basin) yielded post-Villafranchian to Cromerian remains (survey in Radulesco & Samson 1990, 227-228). Remarkable here are *Bison* cf. *schoetensacki* and *Soergelia* cf. *elisabethae* from **Dealul Viilor** (Radulesco & Samson 1965; Samson & Radulesco 1965; Samson 1976), which is younger than Untermassfeld.

Two loess horizons of **Betfia VII/4** with intercalated soil formation (Cromerian I according to Terzea 1996, 538) contained a *Mimomys savini* fauna which is likewise younger than Untermassfeld and includes *Bison schoetensacki*, *Alces latifrons*, and *Mammuthus* cf. *trogontherii*.

A series of sites above the M/B boundary have become known from the Brasov Depression and may clearly be separated from Untermassfeld because of their early Middle Pleistocene faunal elements. To be mentioned here are **Feldioara Cariera** with *Capreolus suessenbornensis*, *Alces latifrons*, *Cervus acoronatus*, *Equus* cf. *mosbachensis*, and *Mammuthus trogontherii*; the similar faunas from **Rotbav-Dealul Tiganilor** ("lower level" and "upper level 1", Cromer IV according to Terzea 1996, 115); and the Elsterian association of **Araci Cariera**, which bears a more continental character because of the presence of *Bison* cf. *priscus*, *Megaloceros savini*, *Coelodonta antiquitatis* ssp., and *Equus* cf. *mosbachensis* (Radulesco et al. 1965; Radulescu & Kovács 1966, 1968; Samson et al. 1969; Radulescu & Samson 1985). The occurrence of *Saiga*, reported for the late Elsterian (Samson & Radulesco 1972, 321ff.) cave infill of **Gura Dobrogei 2**, was in the meantime negated (Radulesco et al. 1998, 286).

In Moldova finds of Late Villafranchian faunas were made at **Chishmikioi** on the Prut. They include *Equus* cf. *stenonis* and *Mammuthus meridionalis* (Ali-Zade et al. 1972, 46ff.; Nikiforova 1982, 27ff.).

A comparatively limited fauna with *Equus* cf. *suessenbornensis* from the upper section of the Matuyama polarity zone is known from **Valeny** (Prut) (Ali-Zade et al. 1972, 13ff.). Large quantities of faunal remains were recovered from the early Middle Pleistocene gravels of the Dnestr at **Tiraspol** (Kolkotova Balka), **Sukleya**, and **Malaeshty** (Nikiforova et al. 1971; Beljaeva & David 1975; Dubrovo 1975; Gromova & Dubrovo 1975; David 1977, 1980; David & Svistun 1981, 11ff.; Nikiforova 1982; David et al. 1990; Croitor 1997). With, e.g., *Praemegaceros verticornis, Cervus acoronatus*, *Equus* cf. *mosbachensis*, and *Mammuthus trogontherii*, the large mammal ranges of these *Mimomys savini* faunas show an age that is markedly younger in comparison with Untermassfeld.

Beginning as early as the first half of the 20th century, a considerable number of mastodon finds and remains of some accompanying species have been described from the Pliocene of Bulgaria. Finds of Late Pleistocene faunas are also known from various localities. Until now, however, Early Pleistocene to early Middle Pleistocene material has scarcely been reported. Spassov & Crégut-Bonnoure (1999) and Spassov (2003) reported on an extensive later Middle Villafranchian large mammal fauna from **Slivnitsa**, west of Sofia, which has to be considered as clearly older than Untermassfeld. It includes *Gazellospira* cf. *torticornis*, *Procamptoceras* cf. *brivatense*, a representative of the *Cervus* s.l. *nestii* group, *Eucladoceros* cf. *tegulensis* (= *senezensis* = *ctenoides*), *Equus* cf. *stenonis*, and *Canis* ex. gr. *etruscus*.

Surveys on the Quaternary faunas of G r e e c e are to be found in Tsoukala (1992) and Koufos & Kostopoulos (1997). A Pliocene to latest Villafranchian age is to be assigned to the collections gathered by Sickenberg near **Polylakos**, **Livakos**, and **Kapetanios** (Haliakmon basin) in the northwest of the country (Steensma 1988; Koufos & Kostopoulos 1997). The material includes the taxa *Leptobos* and *Anancus arvernensis*, which are no longer represented at Untermassfeld.

The rich mammalian fauna from **Gerakarou 1** (Macedonia) has to be placed in the Late Villafranchian (Koufos & Melentis 1983; Koufos 1986, 1992; Kostopoulos & Koufos 1994; Kostopoulos & Athanassiou 1997; Koufos & Kostopoulos 1997; Kostopoulos 1998; Spassov 2003). The chronological distance from Untermassfeld is here demonstrated by, e.g., *Leptobos* cf. *etruscus*, *Eucladoceros tegulensis* (= *senezensis* = *ctenoides*), *Sus strozzii*, *Pliocrocuta perrieri*, and *Canis etruscus*. A collection similar to that from Gerakarou, with *Leptobos* and *Gazellospira* cf. *torticornis*, is known from **Krimni 1** (Macedonia) (Kostopoulos & Koufos 1994; Koufos & Kostopoulos 1997). The material from the **Ravine of Vulgarakis** (Macedonia), which includes *Hippopotamus amphibius antiquus*, is attributed to the latest Villafranchian (Koufos et al. 1989; Koufos 1992; Koufos & Kostopoulos 1997). To judge by the small mammals, the fauna from **Kaiafas** (western Peloponnesus) is Early Biharian. It includes *Alces* cf. *latifrons* and *Canis* cf. *mosbachensis*. Meulen & Kolfschoten (1986, 209, Figs. 2-3) correlate it with Betfia II (see above) and hence this fauna, too, is older than the finds from Untermassfeld.

Relatively close to Untermassfeld chronologically is the fauna from **Apollonia 1** (Macedonia), which was discovered in 1990 (Koufos 1992, 1993; Kostopoulos & Koufos 1994; Martinez-Navarro & Palmqvist 1996; Kostopoulos 1997; Koufos & Kostopoulos 1997; Koufos et al. 1997; Spassov 2003, Croitor & Kostopoulos 2004). The artiodactyl fauna, with the new form *Soergelia brigittae* as well as *Praeovibos* sp., *Pontoceros ambiguus mediterraneus, Bison (Eobison)*, and *Eucladoceros giulii* (Made 1998, 275, 1999, 397), includes a large bovid that for the time being has been referred to *Bison (Eobison)* (Kostopoulos 1997). This species and the occurrence of *Ursus etruscus* and *Canis etruscus* suggest an age greater than that of Untermassfeld.

Chronologically, the fissure infill **Tourkovounia 2** (= Tourkobounia, Athens), with *Macaca sylvanus florentina* (Symeonidis & Zapfe 1977), may approximately be equated with the finds from Untermassfeld on the basis of the small mammals (Maul 2001a, 873). On the other hand, the stratigraphical range of the fissures **Tourkovounia 3-5** extends over a rather long period that possibly reaches into the early Middle Pleistocene (Symeonidis & Vos 1977).

Because the presence of *Praemegaceros verticornis* and *Panthera leo fossilis* was established there (Sickenberg 1976; Koufos & Kostopoulos 1997), **Maratousa** (Megalopolis, Peloponnesus) is younger than Untermassfeld. Meulen & Kolfschoten (1986, 210, figs. 2-3) place the faunule of **Volos** (Thessalia) in the early Galerian. The Pleistocene cave infill of **Petralona** (Chalkidiki) contains mixed remains of two faunas at different age levels. An early Middle Pleistocene fauna could be separated from a late Middle or early Late Pleistocene component and is placed above the M/B boundary (Tsoukala 1989, 1991 with survey of older literature, 1996). The evidence of *Praemegaceros* sp., *Pliocrocuta perrieri*, and *Ursus deningeri* bears out an age younger than that of Untermassfeld.

In T u r k e y a fauna from an open landscape with, e.g., cf. *Leptobos* sp., cf. *Eucladoceros*, *Equus stenonis* ssp., and *Mammuthus meridionalis* is known from late Middle to early Late Villafranchian fluviatile deposits at **Yukari-Sögütönü** (Sickenberg et al. 1975, 44-45, 55). It clearly surpasses Untermassfeld in age. The finds from **Kamisli** (= Merzifon) reported by Sickenberg & Tobien (1971, 60f.) and Sickenberg et al. (1975, 55, 77) should probably likewise be considered as older than Untermassfeld, particularly because of the joint occurrence of *Canis etruscus* and *Pliocrocuta perrieri*. However, according to H. Hemmer (oral comm., February 2002), the canid involved is a Middle Pleistocene wolf.

On the basis of its large mammals (including, e.g., *Equus mosbachensis*), **Dursunlu** (Gülec et al. 1999; Howell et al. 1999) represents a fauna from the late Matuyama polarity zone. Compared to Untermassfeld it is certainly younger. A still little-known Early Palaeolithic cave infill was excavated near **Yarimburgaz**. In this case, too, *Ursus deningeri*, predominant among the bear remains (Tsoukala 1996), is indicative of a younger age.

In the European part of R u s s i a , the Late Villafranchian is characterized by the Psekups Faunal Complex (= middle Akchagylian to early Apsheronian) (Vangengejm et al. 1990, 1991; Titov 1999, 19). The stratotype of the (earlier)

Psekupsian, with, e.g., *Equus stenonis* and a representative of the typical form of *Mammuthus meridionalis*, lies between Saratovskaya and Bakinskaya in the **Psekups** valley near Krasnodar (Alekseeva 1977, 27f.). Limnic deposits at **Mal'zev** and **Georgievsk** in the northern Caucasus likewise produced remains of *Mammuthus meridionalis* (Alekseeva 1977, 26). A skeleton was found in the latter locality (Garutt & Safronov 1965). Finds of *M. meridionalis* are known from **Khutor Azov** and several other occurrences of the so-called "Scythian Clays" on the coast of the Sea of Azov (Lebedeva 1978, 73f.). Older than Untermassfeld is the fossil assemblage of **Semibalki 3** at the Sea of Azov. It includes *Bison (Eobison) tamanensis* (Bajguševa 2000; Bajgusheva et al. 2001).

The type locality of the late Early Pleistocene Taman Faunal Complex of Russia (= middle Apsheronian and earlier part of the late Apsheronian) (Vangengejm et al. 1991) is **Sinyaya Balka** on the northern shore of the Taman peninsula (Dubrovo 1963; regional survey in Bajgusheva et al. 2001). *Mammuthus meridionalis tamanensis* was first described from this site (Dubrovo 1964). Part of the elephantid remains from Sinyaya Balka (the *tamanensis* molars) were attributed to the Jaramillo and would therefore be chronologically close to Untermassfeld. Possibly younger finds from Sinyaya Balka (primitive *trogontherii* molars) might be attributable to the ultimate section of the Matuyama polarity zone (Vangengejm et al. 1991). However, Sher (1999) and Lister et al. (2005) considered contemporaneity of the two morphologies a likely alternative.

Below the M/B boundary, remains of *M. meridionalis tamanensis* (Dubrovo 1964, 90; Dubrovo & Alekseev 1964) as well as *Canis (Xenocyon)* sp. [oral comm. by M. V. Sotnikova (Moscow)] were also recovered near **Margaritovka** (Taganrogskij Saliv, Sea of Azov) (Vangengejm et al. 1991, 47). A fauna with *Equus* cf. *suessenbornensis* is known from **Port Katon** (Sea of Azov) (Bajguševa 1968, 14). Here the results on the small mammals suggested an age similar to that of Untermassfeld (Maul 2001a, 869).

Near the Sea of Azov large mammal remains of the early Middle Pleistocene Tiraspol Faunal Complex (= latest late Apsheronian to Bakuian) (Vangengejm et al. 1991) were also occasionally found. *Mammuthus trogontherii*, for example, was recorded from **Gerasimovka** (Alekseeva 1977, 37-38). A skeleton of the steppe mammoth was recovered from the Kagal'nik Kar'er near the city of **Azov** (Bajguševa & Garutt 1987). This form is also known from the **Kar'er Girej** near Kropotkin and some other sites in the area of the northern Caucasus (Alekseeva 1977, 38). In addition, the Kar'er Girej yielded *Bison* cf. *schoetensacki* and an elaphine deer (David 1977, 7). The abundant finds from layers 6-7a in the northern Caucasian **Treugol'naya Cave** (Karachay-Circassia), with, e.g., cf. *Capreolus suessenbornensis*, *Panthera leo*, *Crocuta crocuta*, and *Ursus deningeri*, demonstrate that their stratigraphical position is higher than that of Untermassfeld. Baryšnikov (1993) correlated the separately maintained Caucasian Urup Faunal Complex with the Tiraspol Faunal Complex.

In the Ukraine the Odessa Faunal Complex (= higher section of the Khaprov Complex according to some authors) is chronologically equated with the Late Villafranchian Psekups Faunal Complex of Russia. The Ukrainian type locality lies on the western shore of the Kuyalnik liman near Odessa (Alekseeva 1977, 23). Particularly from the profile of the Lestniza Trobotti of **Zevakhova Gora,** many large mammal remains, some of which belong to *Equus stenonis* and *Mammuthus meridionalis*, were recovered (listed in Alekseeva 1977, 23; Dubrovo & Kapelist 1979, 88). They are probably not all of the same age. Similar faunas are known from **Chebotareva** on the shore of the Khadzibey liman, as well as from **Bolshaya Kamyshevakha** near Kharkov (Alekseeva 1977, 24-26; Dubrovo & Kapelist 1979, 85).

Finds from the upper horizon of **Kryzanovka** near Odessa, which include an elaphine deer, stenonid equids, and *Mammuthus meridionalis tamanensis*, are attributed to the late Early Pleistocene Taman Faunal Complex (Dubrovo & Kapelist 1979, 92). Near the town of **Nogaisk** on the coast of the Sea of Azov a skeleton of the last-mentioned species was recovered (Garutt 1954). To judge by the arvicolids, the finds from that site are older than the fauna from Untermassfeld (Maul 2001a, 869). The material from **Kairy** is chronologically close to the finds from Nogaisk (Dubrovo & Kapelist 1979, 88-89; Rekovec 1994, 15). It is hard to assess the assemblage from **Dolinskoye** near Odessa. Beside *Bison suchovi* (?= *B. tamanensis*; cf. Sher 1997, 174) it includes *Eucladoceros* and *Mammuthus meridionalis tamanensis*, possibly also *Elephas (Palaeoloxodon) antiquus* (Alekseeva 1967; Ali-Zade et al. 1972, 56ff.; Dubrovo & Kapelist 1979, 87; David & Svistun 1981, 3-4).

Occurrences of the Tiraspol Faunal Complex (see above), which is younger than Untermassfeld, have more often been found in the Ukraine. **Nagornoye** near Odessa, for example, yielded a fauna which includes *Bison schoetensacki, Alces latifrons, Praemegaceros* cf. *verticornis*, *Cervus acoronatus*, *Equus* cf. *mosbachensis*, and *Mammuthus trogontherii* (Alekseeva 1977, 37; David 1977, 5-6). From **Krasnoselka** on the shore of the Kuyalnik liman a similar fauna was described, which likewise includes *Bison schoetensacki*, *Equus mosbachensis*, and *Mammuthus trogontherii* (Alekseeva 1977, 37). **Tikhonovka** near Saporozhe in the southern Ukraine (David 1977, 5; David et al. 1990, 37), with, e.g., *Bison priscus*, *Cervus elaphus*, and *Equus mosbachensis,* is probably attributable to even later in the Middle Pleistocene.

In G e o r g i a the rich faunas found at Dmanisi and Akhalkalaki chronologically 'sandwich' the fossil site of Untermassfeld. Based on several large mammal taxa (the cervids, typical *Mammuthus meridionalis*, a more primitive form of *Megantereon cultridens*, *Pliocrocuta perrieri*, *Ursus etruscus*, etc.), the Late Villafranchian hominid site of **Dmanisi**, where excavations are still going on (L. Gabunia & Vekua 1989, 20ff., 1990, 644f., 1993, 1995; Vekua & L. Gabunia 1989; Vekua 1991, 1995; Bräuer & Schultz 1996; Vekua & Lordkipanidze 1998, 263; Gabunia et al. 1999a, 1999b, 1999–2000, 2000; Vekua et al. 2002; Kiladze 2003), is clearly of greater age in comparison with Untermassfeld. A study of the likewise instructive bovid remains was recently finished (Bukhsianidze 2005). A normally polarized (Olduvai) dolerite with K/Ar ages of 1.8 $\pm$ 0.1 Ma BP (Majsuradze et al. 1989) and 2.0 $\pm$ 0.1 Ma BP (error-weighted average) (Schmincke & Bogaard 1995), underlies the fossiliferous Quaternary succession of Dmanisi.

The assemblages of **Kotsakhuri** (with **Taribana**) in the east and **Tsalka** in the south of the country, each predominantly documented by occasional finds (Vekua et al. 1985; Gabunia & Vekua 1990, 643-644; Vekua 1991, 313-316), are probably likewise chronologically close to Dmanisi (Taribana Faunal Complex) and hence surpass Untermassfeld in age.

The South Georgian fauna from **Akhalkalaki** (Vekua 1962, 1986, 1991, 320ff.; R.-D. Kahlke 1985b; L. Gabunia & Vekua 1989, 42ff., 1990, 645; Hemmer et al. 2001) has to be placed above the Jaramillo polarity subzone and is comparable to Dorn-Dürkheim 3 (see above). Layer 5 of the cave **Kudaro 1** on the southern slope of the Great Caucasus (South Ossetia) contained a Middle Pleistocene warm-stage fauna. Because it includes *Capreolus capreolus, Panthera pardus, Crocuta crocuta*, and *Ursus deningeri* (Baryšnikov 1986, 34ff., 43ff., 1987, 5ff., 1991; Guérin & Barychnikov 1987) it is distinctly younger than Untermassfeld.

From A r m e n i a a (mixed) faunule is known from **Leninakan** (Avakân 1950, 1952, 1959; Alekseeva 1977, 41). It apparently comprises latest Early Pleistocene to Middle Pleistocene elements (stenonid equids, *Mammuthus trogontherii*).

In the western part of A z e r b a i j a n the lagoonal deposits of **Palan-Tyukan** yielded a Late Villafranchian fauna (Sablin 1990; Kuz'mina & Sablin 1991; Sotnikova & Sablin 1993; Vislobokova et al. 1995, 373). Here such taxa as *Sus strozzii*, *Pliocrocuta perrieri*, and *Ursus etruscus* indicate an age greater than that of Untermassfeld.

The **Azykh** Cave in the autonomous region Nagorno-Karabakh produced early Middle Pleistocene faunal remains from layer 6 ("early Acheulean"). These include *Bison* cf. *schoetensacki*, *Cervus acoronatus*, and *Ursus* cf. *deningeri* (Aliev 1969; Baryšnikov 1987, 7-9).

A very important Late Villafranchian fauna is known from eastern K a z a k h s t a n. Sotnikova et al. (1997, 252f.) gave an age estimate of maximally 1.7 Ma BP for the rich assemblage of **Kopaly** (= Aktogai)[26] (layer 7 of the Ili series) (Kožamkulova 1969, 119ff., 1990; Kožamkulova et al. 1987; Kočenov & Kožamkulova 1988, 1990; Tût'kova 1988; Vislobokova et al. 1995, 369, 373). In its macrofauna of Central Asian character, *Leptobos*, *Mammuthus gromovi* (= early form of Eurasian *meridionalis* mammoths), *Canis etruscus*, and other taxa indicate that it is older than Untermassfeld. A North Kazakhian large mammal fauna of rather European to West-Siberian character, recovered near **Pyatoryzsk** and including *Mammuthus meridionalis*, is apparently to be positioned below the Jaramillo polarity subzone (Kožamkulova 1969, 121, 1977, 90; Zylkibaev 1975, 32; Vislobokova et al. 1995, 369, 373).

Of early Middle Pleistocene age and therefore younger than Untermassfeld is the fauna from **Koshkurgan** near Turkestan. It includes *Cervus elaphus* and *Equus* cf. *mosbachensis* (Kožamkulova 1969, 125ff., 1977, 92ff.; David 1977, 10-11) and is correlated with the Tiraspol Faunal Complex.

Some less voluminous material was also found in T u r k m e n i s t a n. Several specimens

[26] In the older literature the place-name Aktogai is found as a synonym for the cited locality of Kopaly, which is on the left bank of the River Charyn. A second profile named Aktogai was opened up on the right bank of the Charyn, 14 km north of Kopaly [kindly communicated orally by M. V. Sotnikova, and by A. E. Dodonov in lit. (both Moscow)]. The latter site, too, yielded a few large mammal remains, but is not mentioned in the present survey.

from **Monzukly** in the palaeo-valley of the Amu Darya demonstrate the presence of *Mammuthus meridionalis* at the base of the late Apsheronian (Dubrovo & Nigarov 1990, 37). In the same area, near **Khudai Dag**, the skull as well as postcranial elements of an early Middle Pleistocene straight-tusked elephant were found, which Dubrovo (1955) described as a new species *Hesperoloxodon turkmenicus* (Dubrovo 1960, 68ff.; Dubrovo & Nigarov 1990, 37-38). The finds from **Tcheleken** in the vicinity of Krasnovodsk (Dubrovo & Nigarov 1990, 38) apparently also include an early Middle Pleistocene component.

In southern Tadzhikistan a late Early Pleistocene fauna was found in reversely magnetized layers of **Lakhuti 2**, a few metres below the M/B boundary [Nikiforova & Vangengejm (eds.) 1988; Sotnikova 1989; Sotnikova & Vislobokova 1990; Sotnikova et al. 1997, 253-254; Forsten & Sharapov 2000]. The fauna includes *Bison* cf. *schoetensacki* and many important carnivore remains, among them *Panthera onca gombaszoegensis*, *Pachycrocuta brevirostris*, *Canis (Xenocyon) lycaonoides*, and *Canis* cf. *mosbachensis*. Comparisons with the Untermassfeld canids are to be found in Sotnikova (2001).

A stratigraphical position similar to that of Lakhuti 2 was attributed to the material from **Tepke 1** in northeastern Kyrgyzstan (Sotnikova et al. 1997, 254). Here the fauna was likewise recovered below the M/B boundary and includes *Bison* sp., a caballoid equid, *Pachycrocuta* cf. *brevirostris*, and *Canis* ex gr. *etruscus* / *mosbachensis*. The early Middle Pleistocene fauna from **Tepke 2** with normal polarization (Brunhes) contained evidence of the elaphine deer *Cervus* cf. *acoronatus* and of *Mammuthus trogontherii trogontherii* (Sotnikova 1997, 255). Both faunas are therefore younger than Untermassfeld. A still younger early Middle Pleistocene age has to be assigned to the finds from the layers 3-5 in the **Sel'-Ungur** Cave in the western part of the country. They contained, e.g., *Panthera* cf. *pardus*, *Hyaena prisca*, *Ursus deningeri*, and *Cuon priscus* (Baryšnikov & Batyrov 1994, 4ff.).

In Syria the fauna from **Latamne** (Hooijer 1961, 1966; Guérin et al. 1993; Tchernov et al. 1994, 329, 332) is notable. Compared to Untermassfeld it turns out to be younger, mainly because of the presence of *Bison* cf. *priscus*, *Dama mesopotamica*, and *Stephanorhinus* cf. *hemitoechus*.

In Israel the chronological framework with respect to the finds from Untermassfeld is formed by the sites of Ubeidiya, Evron, and Gesher Benot Ya'aqov. According to Tchernov (1987, 1988, 191ff., 1999, 93) and Tchernov et al. (1994), the faunal remains from **Ubeidiya** (Tchernov & Guérin 1986; Guérin et al. 1996, 711-712) date to approximately 1.4 Ma BP. Although the geographical vicinity of the African continent (*Hippopotamus gorgops*, early appearance of *Crocuta crocuta*, etc.) and a clearly endemic faunal component here complicate a direct comparison with Untermassfeld, some large mammal taxa of Eurasian origin (*Leptobos, Gazellospira torticornis*, *Sus strozzii*, *Ursus etruscus*) yet suggest that Ubeidiya is older. Within this context, however, the established presence of *Praemegaceros verticornis* (Geraads 1986, 145-150; Guérin et al. 1996, 711) was a surprise. In Central Europe, insofar as known today, this form occurs only above the Jaramillo polarity subzone and is therefore also absent from Untermassfeld. Tchernov (1987, 23) postulated an East Mediterranean centre of origin of the *verticornis* deer.

Tchernov et al. (1994) placed the fauna from **Evron**, which to date comprises only 12 taxa, in a period that lies well before 0.8 Ma BP. There are no opportunities for direct faunistic comparisons with Untermassfeld.

The finds from **Gesher Benot Ya'aqov** (Hooijer 1959, 1960; Tchernov et al. 1994) prove to be younger than Untermassfeld, for instance because of the occurrence of *Dama* cf. *mesopotamica* and *Stephanorhinus hemitoechus* (cf. Latamne). Tchernov et al. (1994, 338) accepted an absolute age of < 0.8 Ma BP. It proved possible to demonstrate the presence of the M/B boundary within the Benot Ya'aqov Formation (Verosub et al. 1998).

Occasional finds of faunal remains of presumably Early Pleistocene age are also known from Saudi Arabia. In the sandy area of **An Nafud**, three assemblages with remains of an animal world of distinctly African character were found, including *Panthera* cf. *onca gombaszoegensis* and *Crocuta crocuta* (Thomas et al. 1998).

8. Importance for faunal history

As already stated, the finds from Untermassfeld illustrate a stage of faunal development which is still little known within the Eurasian area and distinguishes itself from both Late Villafranchian and latest Early Pleistocene to early Middle Pleistocene associations. Considerations that have hitherto been based on a relatively small number of finds per case (cf. Section 7.4.) have mostly made Palaearctic faunal communities with ages between 1.2 and 0.9 Ma BP appear as a simple mix of "typically Villafranchian" and likewise "typically early Middle Pleistocene" (Galerian) elements (cf., e.g., Azzaroli 1995, 312). However, the more thorough evaluation of the first large-scale find complex dated to the said time span, as represented at Untermassfeld, shows that many mammalian groups embody stages of development that lend a very independent character to the fauna as a whole (cf. also R.-D. Kahlke 2000a, 2004):

At Untermassfeld the replacement of the *Eobison* group by *Bison* s. str. is found to be accomplished, without *Bison schoetensacki* or *B. priscus* having yet made their appearance. Insofar as known today, *Capreolus cusanoides* has neither been recovered from Villafranchian, nor from early Middle Pleistocene horizons. Both *Cervus* s.l. *nestii vallonnetensis* and *Eucladoceros giulii* represent independent forms within evolutionary lineages that originated during the Late Villafranchian. Above the M/B boundary these two cervids are superseded by early Middle Pleistocene species.

The evolutionary level of *Hippopotamus amphibius antiquus* from Untermassfeld may be discriminated (mainly on the basis of its skull morphology) both from those of the Late Villafranchian, and from those of certain early Middle Pleistocene age. The Villafranchian *Sus strozzii* had already been replaced by an early form of *Sus scrofa*.

The Untermassfeld individuals of *Stephanorhinus etruscus* (sensu H.-D. Kahlke 2001b; see above) tend to distinguish themselves by their body dimensions, both with respect to Late Villafranchian material from the type region and in relation to remains from Central-European early Middle Pleistocene find complexes.

The appearance of *Acinonyx pardinensis pleistocaenicus* in Untermassfeld signals a trans-Palaearctic gene flow from the east or otherwise an immigration of Asian cheetahs into Europe (Hemmer 2000). *Megantereon cultridens adroveri* represents the most specialized stage of this type of dirktooth cat. In Europe, finds younger than those from Untermassfeld have not been made.

Pachycrocuta brevirostris is the only hyaena represented in the Untermassfeld faunal system. Indications of *Chasmaporthetes lunensis*, *Pliocrocuta perrieri*, and / or *Crocuta crocuta* ssp. are absent. *Ursus* cf. *dolinensis* from Untermassfeld stands out against older *U. etruscus* populations, particularly in the more appreciable dimensions of its body and dentition, and also in the incipient reduction of the premolars. The combination of characters shows some similarities to the younger forms from the type locality of Atapuerca-Gran Dolina.

Within the canid fauna, too, a situation that differs from older conditions has been documented. *Canis (Xenocyon)* ex gr. *falconeri*, *C. etruscus*, and *C. arnensis* no longer occur. The species found instead – *C. (Xenocyon) lycaonoides* and *C. mosbachensis* – show stages of development that to a major degree are morphologically separable from those of early Middle Pleistocene populations.

In 1988 Azzaroli et al. (p. 89) placed the transition from Villafranchian to Galerian faunas – the so-called "end-Villafranchian" dispersal event – chronologically close to the Jaramillo polarity subzone. The idea that a somewhat longer time span was needed for this faunal turnover was formulated later (Azzaroli 1995, 311ff.). Kolfschoten (1994, 27) and also Roebroeks & Kolfschoten (1995, 303) contradicted the concept of a short-lived, "abrupt" change from Villafranchian to Galerian faunas, and estimated a period of more than 500 ka for the duration of this process. Agusti & Moyà-Solà (1998) generally rejected the notion of an "end-Villafranchian" event because they perceived that the Early Pleistocene renewal of the fauna had already started at a time close to the Late Pliocene Olduvai polarity subzone, with the earliest dispersal of Asian or Beringian faunal elements (e.g., *Allophaiomys pliocaenicus*, *Lemmus*, Ovibovini) into Europe.

Against the background of this discussion a redefinition of the biostratigraphical units of the Villafranchian (Azzaroli et al. 1988) and the Galerian (Sala et al. 1992) becomes necessary, and with that a more exact subdivision of the Early Pleistocene to the early Middle Pleistocene. Suitable possibilities for a solution to this problem, that can only be reached by convention, were listed by Caloi & Palombo (1995, 394ff.): A shift of the upper boundary of the Villafranchian to the M/B boundary or a shift of the lower boundary of the Galerian into the late Early Pleistocene (Jaramillo), or otherwise the introduction of a "Protogalerian" (with various conceivable stratigraphical ranges) between the two units.

As already stated above, the Untermassfeld

finds document a stage of faunal development still imperfectly known in the Eurasian area, and nonetheless distinct from both the Late Villafranchian and the latest Early Pleistocene to early Middle Pleistocene (Galerian) mammal associations. Thanks to its largely independent character, the Untermassfeld large mammal fauna – the most complete mammal complex so far known in Europe, but also in the whole Western Palaearctic, from 1.2 to 0.9 Ma BP – supports the idea of a chronostratigraphical unit to be included between the Villafranchian (Azzaroli et al. 1988) and the Galerian (Sala et al. 1992) biochrons. Bourdier already in 1961 proposed such a new formal unit, the "Epi-Villafranchien", based on the geological sections of two French sites, e.g. Durfort and Saint-Cosme (Bourdier 1961, 245-249). He meticulously described some time-diagnostic proxies of the two sections, such as their pollen and mollusc content; he also reported the occurrence of some vertebrate remains. Bourdier had created what we now call a biochron. Bourdier did not support his proposal with a convincingly rich fossil record (see above) and so it was backed only by a minority of vertebrate palaeontologists. From the 1990s onwards, when the Untermassfeld fossil assemblage grew richer, the Epivillafranchian regained consideration (e.g., R.-D. Kahlke 1991-2004, Unger & R.-D. Kahlke 1995, Hemmer 2000, 2001).

If the Epivillafranchian can be resumed as a biochron *sensu* Lindsay (1990: 5-6; following Hedberg 1976) and Steininger & Piller (1999), the application of the term should base on an emended definition. The increased palaeontological data for the different large mammal taxa discussed above are proposed to be taken into consideration. Especially the association *Bison menneri* / *Eucladoceros giulii* / *Cervus* s.l. *nestii vallonnetensis* / *Capreolus cusanoides* / *Stephanorhinus etrucus/hundsheimensis*, which could be named *Bison menneri-Eucladoceros giulii*-assemblage, would provide a suitable basis to improve the definition of the Epivillafranchian biochron ("*Bison menneri-Eucladoceros giulii*-assemblage chron"). The listed taxa of the given assemblage were extensively described and figured by Sher (1997) and H.-D. Kahlke (1997, 2001a, 2001b) from Untermassfeld (Germany). In Europe (and also in large parts of Asia) typical elements of this biochron, although less significant stratigraphically, are *Alces carnutorum*, *Panthera onca gombaszoegensis*, *Acinonyx pardinensis pleistocaenicus*, *Megantereon cultridens adroveri*, *Puma pardoides*, *Pachycrocuta brevirostris*, *Ursus* cf. *dolinensis*, *Canis (Xenocyon) lycaonoides*, and *Canis mosbachensis*. During the time span corresponding to the Epivillafranchian, some of them show distinct evolutionary stages (see above). The fossil assemblage from Untermassfeld, which revives Bourdier's biochron, can be considered a principal reference material for the large mammal fauna of the Epivillafranchian.

Future research will show whether an Epivillafranchian zone, thus defined at temperate European latitudes, is going to be practicable, i.e., supraregionally usable and correlatable. Not only should comparisons with other stratigraphical schemes be made, but it is also necessary that find complexes within this time bracket from other biogeographical regions are taken into consideration.

It is obvious that the process of an Epivillafranchian faunal renewal, as it is now becoming more apparent, must be viewed in a global context. Already Azzaroli (1995, 312ff.) showed that, irrespective of several marked warm stages, indications of a cooling trend and an increase of continental conditions between 1.2 and 0.9 Ma BP may be observed in the entire Eurasian area. Analyses of loess deposition and pollen sequences, in particular, suggest rather serious changes in regional ecological conditions. At that time, subtropical Africa was likewise characterized by significant changes in climatic variability (deMenocal & Bloemendal 1995). The $\delta^{18}O$ values of Ocean Drilling Program (ODP) site 677 (MIS 36-22) show an increase in amplitude, even though the borehole is situated near the equator in the eastern Pacific (Shackleton 1995, fig. 17. 1.). Something similar goes for ODP site 659 off NW Africa and site 607 in the central part of the northern Atlantic (Tiedemann et al. 1994, fig. 3).

The Epivillafranchian was therefore characterized by an increased climatic variability that in the end also created the ecological preconditions for the dispersal of early hominids in Europe.

Summary

Since 1978, an excavation covering an area of 584 m², and in places extending as much as 5.0 m below the surface, has been carried out near Untermassfeld (southern Thuringia, Central Germany). A total of 9,400 determinable remains of larger vertebrates have been found, together with 3,000 small vertebrate specimens, all of which are included in an interdisciplinary study. Thus far, a total of 105 taxa have been identified from the site (Gastropoda: 36, Osteichthyes: 3, Amphibia: 8, Reptilia: 6, Aves: 8, Mammalia: 44).

All larger mammal remains found during the planned excavations up to 1995 are included on the enclosed excavation maps. The state of preservation, the large- and smaller-scale distribution patterns of the finds, and bone modifications and bone spectra, are described. In addition to the determination of the minimum number of individuals and the individual age distribution of the larger mammalian taxa, reconstructions of assemblages of bones and individuals complete the picture.

The lithological features of the fossil-bearing horizons at Untermassfeld indicate a repeated succession of roughly similar high flood events. Several of these events carried carcasses, both complete and fragmented, into the area where the recent excavations have taken place. The accumulation of all skeletal material occurred in a similar position under the lee of a clastic mudflow fan, which interfingers with the section from the east. It was calculated that the finds which enriched the concentrated lagerstätte originated from a potentially flooded area of approximately 94 km². Upstream of the excavation site a narrow part of the valley caused strong turbulence as well as a rapid rise of the water level, which propagated from there against the set of the current, accentuating the catastrophic impact on the fauna.

Approximately 50% of the larger mammal individuals (MNI = 313) were shown to be direct victims of flood events. The range, number of individuals, and age structure of the preserved vertebrate remains had been primarily determined by the population density of the different species within the Early Pleistocene biotope, their specific behavioural patterns and their locomotive abilities. It is also possible that the positions of sleeping localities were influential, as were the weight and transportability of the fresh carcasses.

The taphonomical situation within the fossiliferous layers of the Lower Fluviatile Sands (flooded surfaces) indicates that the deposition of necromass occurred in a low-energy environment. In contrast to this, the accumulation of larger amounts of mostly disarticulated and partly fragmented skeletal elements in the Upper Fluviatile Sands (Channel Infill) indicates that accumulation here began just after the climax of a pronounced high flood event. The decomposition of the Untermassfeld skeletal material was accelerated and modified by the diverse carnivore community, particularly by *Pachycrocuta brevirostris*. Coprolite nests, together with the spatial distribution of small mammal gnaw marks, indicate the intermittent drying-out of sediment surfaces before they were finally covered.

The large number of antlers connected to skull fragments indicates that the accumulation of necromass was restricted to the winter half-year. The faunal remains recovered near Untermassfeld document the development of a fossil site that formed during a geologically short period, in other words, during a biostratigraphically irrelevant time span.

Available geological, taphonomical, and palaeozoological-ecological data together provide a detailed picture of the landscape that existed whilst the Untermassfeld fossil site was evolving. The broad, flat-bottomed valley of the Early Pleistocene Werra contained a dynamic river. Protected areas of the river were characterized by open and clear, low-energy conditions. Within the floodplain there were oxbow lakes with stagnant water and increasing sedimentation, water meadows, permanently wet biotopes with thicket-like vegetation, in addition to stunted woodland in areas susceptible to flooding. Within the valley, open mixed woodland also developed. The Muschelkalk slopes in the vicinity of the site indicate biotopes with low-diversity vegetation comprising trees and shrubs, adjacent to open areas. Park-like landscapes and, in many cases, biotopes with few or no trees, developed on the flanking higher ground.

The composition of the Untermassfeld fauna and the picture of the landscape deduced from it indicate, without doubt, warm climatic conditions. The fossil record predominantly contains temperate to thermophilous elements, which are by varying degrees associated with a suite of biotopes. Higher summer temperatures than those presently experienced in this region can be deduced, as well as mild winters with minimum air temperatures a few degrees below 0°C. A regional climate developed that lacked extreme variations in daily and yearly temperatures because of moderating Atlantic influence.

A clear chronological distinction between the Untermassfeld mammal association and older faunas of the European Villafranchian is illustrated by *Bison menneri*, *Alces carnutorum*, *Cervus* s.l. *nestii vallonnetensis*, *Eucladoceros giulii*, *Hippopotamus amphibius antiquus*, *Stephanorhinus etruscus*, *Panthera onca gombaszoegensis*, *Acinonyx pardinensis pleistocaenicus*, *Megantereon cultridens adroveri*, *Ursus* cf. *dolinensis*, *Canis (Xenocyon) lycaonoides*, *Canis mosbachensis*, and the micromammal fauna. From a faunistic point of view, a maximum age of < 1.2 Ma BP is determined for the Untermassfeld site.

A corresponding minimum age is fixed by utilising *Bison menneri*, *Capreolus cusanoides*, *Cervus* s.l. *nestii vallonnetensis*, *Hippopotamus amphibius antiquus*, *Stephanorhinus etruscus*, *Megantereon cultridens adroveri*, *Homotherium crenatidens*, *Canis (Xenocyon) lycaonoides*, and *Canis mosbachensis*, as well as the micromammals. Comparison of the Untermassfeld fauna with faunas from well-dated younger fossil horizons elsewhere in Europe leads to the conclusion that the current site has a minimum age of approximately 0.9 Ma BP. It is therefore clear that the magnetic boundary identified immediately below the bone-bearing layers must be the base of the Early Pleistocene Jaramillo subzone. Consequently, an absolute age of slightly more than one million years can be ascribed to the Untermassfeld faunal remains.

To illustrate the stratigraphical position of the Untermassfeld fauna, important comparable larger mammal associations from some 180 localities, ranging from Europe to Middle Asia and the Near East, were grouped according to age. It becomes clear that the Untermassfeld fauna marks out a faunistic level which is relatively poorly known in Eurasia and differs significantly from both Late Villafranchian and latest Early Pleistocene to early Middle Pleistocene associations.

European late Early Pleistocene faunal assemblages do not appear as mixtures of "older" (Villafranchian) and "younger" (Galerian) elements. The largely independent character of the Untermassfeld fauna supports the idea of a separate chronostratigraphical unit wedged between the Villafranchian and the Galerian biochrons. This unit was proposed and named the "Epi-Villafranchien" by Bourdier, already in 1961. Should the Epivillafranchian be resumed, Untermassfeld's *Bison menneri* / *Eucladoceros giulii* / *Cervus* s.l. *nestii vallonnetensis* / *Capreolus cusanoides* / *Stephanorhinus etruscus*/*hundsheimensis* association (possibly identified as the "*Bison menneri*-*Eucladoceros giulii*-assemblage") would most suitably represent the biochron.

The Epivillafranchian was characterized by an increased climatic instability, which ultimately provided the ecological background to the dispersal of early hominids in Europe.

References

Abbazzi, L. 2004: Remarks on the validity of the generic name *Praemegaceros* Portis 1920, and an overview on *Praemegaceros* species in Italy. Rend. Fis. Acc. Lincei, s. 9, 15, 115–132, Roma.

Abbazzi, L., Benvenuti, M., Rook, L. & Masini, F. 1995: Biochronology of the Mugello intermontane basin (Northern Apennines, Italy). Il Quaternario 8, 1, 5-10, Verona.

Aguirre, E. (ed.) 1998: Atapuerca y la evolución humana. 453 p., Fundación Ramón Areces, Madrid.

Aguirre, E. & Morales, J. 1990: Villafranchian Faunal Record of Spain. Quartärpaläontologie 8, 7-11, Berlin.

Agusti, J. 1991: The *Allophaiomys* Complex in Southern Europe. Géobios 25, 1, 133-144, Villeurbanne.

Agusti, J. & Moyà-Solà, S. 1991: Les faunes de mammifères du Pléistocène inférieur et moyen de l'Espagne: implications biostratigraphiques. L'Anthropologie 95, 4, 753-764, Paris.

1992: Mammalian dispersal events in the Spanish Pleistocene. In: Koenigswald, W. v. & Werdelin, L. (eds.): Mammalian Migration and Dispersal Events in the European Quaternary, Courier Forschungsinst. Senckenberg 153, 69-77, Frankfurt a. M.

1998: The Early Pleistocene mammal turnover in Spain: evidence against an «End-Villafranchian» event. In: Kolfschoten, T. v. & Gibbard, P. L. (eds.): The Dawn of the Quaternary, Proceedings of the SEQS-EuroMam symposium 1996, Meded. Nederl. Inst. Toegepaste Geowetenschappen TNO 60, 513-519, Haarlem.

Agusti, J., Moyà-Solà, S. & Pons-Moyà, J. 1987: La sucesión de Mamiferos en el Pleistoceno inferior de Europa: proposición de una nueva escala bioestratigráfica. Paleont. i Evol., Mem. Esp. 1, 287-295, Sabadell.

Agusti, J. & Oms, O. 1998: The Middle Pleistocene Cullar area sites. In: Agusti, J., Oms, O. & Martin-Suárez, E. (eds.): The Plio-Pleistocene Vertebrate Succession of the Guadix-Baza Basin (SE Spain), Field Guide, 31-32, European Quaternary Mammal Research Association (INQUA-SEQS), Granada.

Agusti, J., Oms, O. & Pares, J. M. 1999: Calibration of the Early-Middle Pleistocene transition in the continental beds of the Guadix-Baza Basin (SE Spain). Quaternary Sci. Rev. 18, 1409-1417, Oxford.

Alberdi, M. T., Azanza, B., Cerdeño, E. & Prado, J. L. 1997: Similarity relationship between Mammal faunas and biochronology from Latest Miocene to Pleistocene in the Western Mediterranean area. Eclogae geol. Helv. 90, 115-132, Basel.

Alberdi, M.-T., Caloi, L. & Palombo, M.-R. 1988: The Quaternary fauna of Venosa: Equids. Bull. Mus. Anthrop. préhist. Monaco 31, 5-39, Monaco.

Alberdi, M. T. & Sánchez-Chillón, B. 1998: The Cúllar-Baza 1 site. In: Agusti, J., Oms, O. & Martin-Suárez, E. (eds.): The Plio-Pleistocene Vertebrate Succession of the Guadix-Baza Basin (SE Spain), Field Guide, 33-35, European Quaternary Mammal Research Association (INQUA-SEQS), Granada.

Alekseeva, L. I. 1967: K istorii podsemejstva Bovinae v eoplejstocene Evropejskoj časti SSSR. In: Burgelâ, N. K., Negadaev-Nikonov, K. N., Polev, P. V., Safarov, E. I. & Edel'stejn, A. A. (eds.): Paleontologiâ, geologiâ i poleznye iskopaemye Moldavii, Paleontologiâ i stratigrafiâ, 125-142, Akad. Nauk Moldavskoy SSR, Kišinev (in Russian).

1977: Teriofauna rannego antropogena vostočnoj Evropy. 214 p., Nauka, Moskva (in Russian).

Aliev, S. D. 1969: Fauna Azyhskoj paleolitičeskoj stoânki. Avtoreferat dissertazii, 29 p., Akad. Nauk Azerbajdzanskoj SSR, Sekt. Paleobiol., Baku (in Russian).

Ali-Zade, A. A., Alizade, K. A., Aleskerov, D. A., Buleishvili, D. A., Vekua, A. K., Konstantinova, N. A., Lebedeva, N. A., Negadaev-Nikonov, K. N., Nikiforova, K. V., Pevzner, M. A., Khubka, A. N., Chepalyga, A. L. & Chernyakhovsky, A. G. 1972: Guidebook. Excursions in Moldavia, Georgia, Azerbaijan. May - June, 1972. 104 p., INQUA Subcommission on the Pliocene-Quaternary Boundary, IUGS Subcommission on Neogene Stratigraphy, Moscow.

Ambrosetti, P., Bartolomei, G., Giuli, C. de, Ficcareli, G. & Torre, D. 1979: La breccia ossifera di Slivia (Aurisina - Sistiana) nel Carso di Trieste. Boll. Soc. Paleont. Ital. 18, 2, 207-220, Modena.

Ambrosetti, P., Basilici, G., Capasso Barbato, L., Carboni, M. G., Stefano, G. di, Esu, D., Gliozzi, E., Petronio, C., Sardella, R. & Squazzini, E. 1995: Il Pleistocene Inferiore nel ramo sud-occidentale del Bacino Tiberino (Umbria): Aspetti litostratigrafici e biostratigrafici. Il Quaternario 8, 1, 19-36, Verona.

Aouadi, N. & Bonifay, M.-F. 1997-1998: Études paléontologique et taphonomique de restes de proboscidiens (Ceyssaguet, Haute-Loire). Bull. Mus. Anthrop. préhist. Monaco 39, 17-27, Monaco.

Arribas, A., Baeza, E., Bermúdez, D., Blanco, S., Durán, J. J., Garrido, G., Gumiel, J. C., Hernández, R., Soria, J. M. & Viseras, C. 2004: Nuevos registros paleontológicos de grandes mamíferos en la Cuenca de Guadix-Baza (Granada): aportaciones del Proyecto Fonelas al conocimiento sobre las faunas continentales del Plioceno-Pleistoceno europeo. Bol. Geol. Minero 115, 3, 567-582, Madrid.

Arribas, A. & Palmqvist, P. 1998: Taphonomy and palaeoecology of an assemblage of large mammals: Hyaenid activity in the Lower Pleistocene site at Venta Micena (Orce, Guadix-Baza Basin, Granada, Spain). Géobios 31, 3, suppl., 3-47, Villeurbanne.

Arribas, A., Riquelme, J. A., Palmqvist, P., Garrido, G., Vernández, R., Laplana, C., Soria, J. M., Viseras, C., Durán, J. J., Gumiel, P., Robles, F., López-Martínez, J. & Carrión, J. 2001: Un nuevo yacimiento de grandes mamíferos villafranquienses en la Cuenca de Guadix-Baza (Granada): Fonelas P-1, primer registro de una fauna próxima al limite Plio-Pleistoceno en la Península Ibérica. Bol. Geol. Minero 112, 4, 3-34, Madrid.

Ashwin, T. & Stuart, T. 1996: The West Runton Elephant. Current Archaeol. 149, 164-168, London.

Avakân, L. A. 1950: Ostatki *Elephas trogontherii* Pohl. Naučn. trudy Erevanskogo Gos. univ. im. V. M. Molotova, 149-161 + 3 pls., Erevan (in Russian, Georgian summary).

1952: Ostatki *Equus stenonis* iz četvertičnyh otloženij Armenii. Naučn. trudy Erevanskogo Gos. univ. im. V. M. Molotova 37, ser. geol. nauk, 109-117, Erevan (in Russian, Georgian summary).

1959: Četvertičnoe iskopaemye mlekopitaûšie Armenii. 72 p. + XII pls., Izd. AN Armânskoj SSR, Erevan (in Russian).

Azzaroli, A. 1977: The Villafranchian Stage in Italy and the Plio-Pleistocene Boundary. Giorn. Geol. 41, 1-2, 61-79, Roma.

1992: The cervid genus *Pseudodama* n. g. in the Villafranchian of Tuscany. Palaeontograph. Ital. 79, 1-14, Pisa.

1995: The "Elephant-*Equus*" and the "End-Villafranchian" Events in Eurasia. In: Vrba, E. S., Denton, G. H., Partridge, T. C. & Burckle, L. H. (eds.): Paleoclimate and Evolution, with Emphasis on Human Origins, 311-318, Yale Univ. Press, New Haven, London.

Azzaroli, A., Giuli, C. de, Ficcarelli, G. & Torre, D. 1988: Late Pliocene to early mid-Pleistocene mammals in Eurasia: faunal succession and dispersal events. Palaeogeogr., Palaeoclim., Palaeoecol. 66, 77-100, Amsterdam etc.

Azzaroli, A. & Mazza, P. 1993: Large early Pleistocene deer from Pietrafitta lignite mine, Central Italy. Palaeontograph. Ital. 80, 1-24, Pisa.

Bajguševa, V. S. 1968: Paleontologičeskaâ harakteristika po faune mlekopitaûših osnovnyh verhnepliocenovyh razrezov severo-vostočnogo Priazov'â (Livenzovka, Port-Katon). Avtoreferat dissertazii, 18 p., Rostovskij gosudarstvennyj univ., Rostov-na-Dony (in Russian).

2000: Novye dannye o tamanskom faunističeskom komplekse iz raskopka u sela Semibalki (Priazov'e). In: Istoriko-arheologičeskie issledovaniâ v Azove i na Niznem Dony v 1998 g., 27-57, Ministerstvo kul'tury Rostovskoj oblasti, Azovskij kraevedčeskij muzej, Azov (in Russian).

Bajguševa, V. S. & Garutt, V. E. 1987: Skelet stepnogo slona *Archidiskodon trogontherii* (Pohlig, 1885) iz severo-vostočnogo Priazov'â. In: Kuz'mina, I. E. & Baryšnikov, G. F. (eds.): Plejstocenovye mlekopitaûšie severnoj Evrazii, Akad. Nauk SSSR, Trudy Zool. Inst. 168, 21-37, Leningrad (in Russian, English summary).

Bajgusheva, V. S., Titov, V. V. & Tesakov, A. S. 2001: The sequence of Plio-Pleistocene mammal faunas from the south Russian Plain (the Azov Region). Boll. Soc. Paleont. Ital 40, 2,

133-138, Modena.

Bartolomei, G., Chaline, J., Fejfar, O., Jánossy, D., Jeannet, M., Koenigswald, W. v. & Kowalski, K. 1975: *Pliomys lenki* (HELLER 1930) (Rodentia, Mammalia) en Europe. Acta zool. carcov. 20, 10, 393-467, Kraków.

Baryšnikov, G. F. 1986: Plejstocennovye volki (rod *Canis*) Bol'sogo Kavkaza. In: Verešagin, N. K. & Kuz'mina, I. E. (eds.): Mlekopitaûšie četvertičnoj fauny SSSR, Akad. Nauk SSSR, Trudy Zool. Inst. 149, 33-52, Leningrad (in Russian, English summary).

1987: Mlekopitaûšie Kavkaza v epohu rannego paleolita. In: Kuz'mina, I. E. & Baryšnikov, G. F. (eds.): Plejstocenovye mlekopitaûšie severnoj Evrazii, Akad. Nauk SSSR, Trudy Zool. Inst. 168, 3-20, Leningrad (in Russian, English summary).

1991: *Ursus mediterraneus* v plejstocene Kvkaza i zamečaniâ po istorii melkih medvedej Evrazii. In: Baryšnikov, G. F. & Kuz'mina, I. E. (eds.): Paleoteriologičeskie issledovaniâ fauny SSSR, Akad. Nauk SSSR, Trudy Zool. Inst. 238, 3-60, Sankt-Peterburg (in Russian, English summary).

1993: Krupnye mlekopitaûšie ašels'koj stoânki v pešere Treugol'naâ na Severnom Kavkaze. In: Baryšnikov, G. F. & Kuz'mina, I. E. (eds.): Materialy po mezozojskoj i kajnozojskoj istorii nazemnyh pozvonočnyh, Ross. Akad. Nauk, Trudy Zool. Inst. 249, 3-47, Sankt-Peterburg (in Russian, English summary).

Baryšnikov, G. F. & Batyrov, B. H. 1994: Sredne-plejstocenovye hisnye mlekopitaûšie (Carnivora, Mammalia) Srednej Azii. In: Kuz'mina, I. E. & Baryšnikov, G. F. (eds.): Četvertičnaâ fauna servernoj Evrazii, Ross. Akad. Nauk, Trudy Zool. Inst. 256, 3-43, Sankt-Peterburg (in Russian, English summary).

Bassinot, F. C., Labeyrie, L. D., Vincent, E., Quidelleur, X., Shackleton, N. J. & Lancelot, Y. 1994: The astronomical theory of climate and the age of the Brunhes-Matuyama magnetic reversal. Earth and Planetary Science Letters 126, 91-108, New York etc.

Beden, M., Bonis, L. de, Brunet, M. & Tournepiche, J.-F. 1984: Première découverte d'un félin machairodonte dans le Pléistocène moyen des Charentes. C. R. Acad. Sci. Paris 298, Sér. II, 6, 241-244, Paris.

Behrensmeyer, A. K. 1988: Vertebrate preservation in fluvial channels. Palaeogeogr., Palaeoclim., Palaeoecol. 63, 183-199, Amsterdam etc.

Beljaeva, E. & David, A. I. 1975: Die Nashörner (Rhinocerotidae) aus dem Faunenkomplex von Tiraspol. Quartärpaläontologie 1, 157-175, Berlin.

Berggren, W. A., Kent, D. V., Swisher, C. C. & Aubry, M.-P. 1995: A Revised Cenozoic Geochronology and Chronostratigraphy. In: Berggren, W. A., Kent, D. V., Aubry, M.-P. & Hardenbol, J. (eds.): Geochronology, Time Scales and Global Stratigraphic Correlation, Soc. Sediment. Geol., Spec. Publ. 54, 131-212, Tulsa.

Bergh, G. D. v. d., Sondaar, P. Y., Vos, J. de & Aziz, F. 1996: The proboscideans of the South-East Asian islands. In: Shoshani, J. & Tassy, P. (eds.): The Proboscidea, Evolution and Palaeoecology of Elephants and Their Relatives, 240-248, Oxford Univ. Press, Oxford etc.

Berinkey, L. 1959: Early Pleistocene Fish fossils from the Üröm quarry. Ann. Hist.-Nat. Mus. nation. Hungar. 51, 105-112, Budapest.

Bermúdez de Castro, J. M., Arsuaga, J. L., Carbonell, E., Rosas, A., Martinez, I. & Mosquera, M. 1997: A Hominid from the Lower Pleistocene of Atapuerca, Spain: Possible Ancestor to Neandertals and Modern Humans. Science 276, 1392-1395, Washington.

Binford, L. R. 1981: Bones. Ancient Man and Modern Myths. 320 p., Acad. Press, Orlando etc.

Bishop, M. J. 1982: The mammal fauna of the early Middle Pleistocene cavern infill site of Westbury-sub-Mendip, Somerset. Spec. Pap. Palaeont. 28, 1-108, London.

Böhme, G. 1996: Fischreste aus dem Eem-Interglazial von Schönfeld bei Calau, Niederlausitz. In: Striegler, R. & Striegler, U. (eds.): Eem von Schönfeld II, Naturlandschaft in der Niederlausitz, special issue, 9-48, Cottbus.

1997a: Die Amphibienreste aus dem Unterpleistozän von Untermaßfeld. In: Kahlke, R.-D. (ed.): Das Pleistozän von Untermaßfeld bei Meiningen (Thüringen), Part 1, Monogr. Röm.-Germ. Zentralmus. Mainz 40, 1, 81-87, Halbelt, Bonn.

1997b: Fossile Fischfaunen aus dem jüngeren

Känozoikum Deutschlands. Quartär 47/48, 113-138, Saarbrücken.

Bolomey, A. 1965: Die Fauna zweier villafrankischer Fundstellen in Rumänien. Ber. geol. Ges. DDR 10, 1, 77-88, Berlin.

Bonifay, E. & Mergoil, J. 1988: Les maars de Soleilhac (Blanzac, Haute-Loire) et leurs séries volcano-sédimentaires. C. R. Acad. Sci. Paris 307, Sér. II, 1561-1566, Paris.

Bonifay, M.-F. 1971: Carnivores quaternaires du Sud-Est de la France. Mém. Mus. natn. Hist. nat., N. S., Sér. C, 21, 2, 43-377, Paris.

1981: Les *Praemegaceros* du Pléistocène moyen de la Grotte de l'Escale à Saint-Estéve-Janson (Bouches-du-Rhone): leur intérêt dans le contexte biostratigraphique européen. Bull. Assoc. Franc. Ét. Quatern. 7-8, 3/4, 109-120, Paris.

1986: Intérêt des études taphonomiques au Pléistocène ancien: Soleilhac et Ceyssaguet (Blanzac, Haute Loire). Bull. Mus. natn. Hist. nat., 4e sér., 8, C, 2, 269-281, Paris.

1988: L'autre préhistoire... Remarques sur les faunes contemporaines du très vieux Paléolithique fançais. Archaeozoologia 2, 1, 2, 181-188, Grenoble.

1990: Relations between paleoclimatology and plio-pleistocene biostratigraphic data in West European countries. In: Lindsay, E. H., Fahlbusch, V. & Mein, P. (eds.): European Neogene Mammal Chronology, NATO ASI Ser. A, Life Sci. 180, 475-485, Plenum, New York, London.

1992: Apport des grandes faunes villafranchiennes du Massif Central français dans le cadre général du Plio-Pléistocène. In: 117è Congr. nat. soc. sav., 55-64, Clermont-Ferrand.

1993: Paleoclimatic correlations between Western european continental and marine areas. Contribution of big villafranchian fauna. Ciéncias da Terra (UNL) 12, 111-115, Lisboa.

Bonifay, M.-F., Bonifay, E. & Malaterre, J.-F. 1984: Le site Pléistocène inférieur de Ceyssaguet (Haute-Loire). Résultats préliminaires. In: 10è Réunion Annuelle des Sciences de la Terre, 73, Soc. Géol. Franc., Bordeaux.

Bonifay, M.-F. & Brugal, J.-P. 1999: Biogéographie et Biostratigraphie des grandes faunes quaternaires en Europe du sud: apport des gisements français. In: Gibert, J., Sánchez, F., Gibert, L. & Ribot, F. (eds.): Los Hominidos y su entorno en el Pleistoceno inferior y medio de Eurasia, Actas del Congreso Internacional de Paleontologia Humana, Orce 1995, 557-566, Museo Prehistoria y Paleontologia «J. Gibert», Orce.

Bosinski, G. 1995: The earliest occupation of Europe: Western Central Europe. In: Roebroeks, W. & Kolfschoten, T. v. (eds.): The Earliest Occupation of Europe, Proceedings of the European Science Foundation Workshop at Tautavel (France), 1993, 103-128, Univ. Leiden, Leiden.

Bourdier, F. 1961: Le Bassin du Rhône au Quaternaire. Géologie et préhistoire. Tome I. Texte. 364 p., C. N. R. S., Paris.

Bowen, D. Q. & Sykes, G. A. 1994: How old is 'Boxgrove man'?, Roberts replies. Nature 371, 751, London.

Bräuer, G. & Schultz, M. 1996: The morphological affinities of the Plio-Pleistocene mandible from Dmanisi, Georgia. J. Human Evol. 30, 445-481, London etc.

Breda, M. & Marchetti, M. 2005: Systematical and biochronological review of Plio-Pleistocene Alceini (Cervidae; Mammalia) from Eurasia. Quaternary Sci. Rev. 24, 5-6, 775-805, Oxford.

Breda, M., Pini, R. & Ravazzi, C. 2005: The palaeoenvironment of *Cervalces latifrons* (Johnson, 1874) from Fornaci di Ranica (late Early Pleistocene, Northern Italy). Palaeogeogr., Palaeoclim., Palaeoecol. 216, 1-2, 99-118, Amsterdam etc.

Brochet, G., Chaline, J. & Poplin, F. 1983: Courterolles (Yonne), une faune interglaciaire a hippopotame (Waalien?) et une microfaune steppique a *Allophaiomys* (Ménapien ?) du Pléistocène inférieur. Mém. Soc. Préhist. France 16, 15-18, Paris.

Brugal, J.-P. 1985: Le *Bos primigenius* Boj., 1827 du Pléistocène moyen des grottes de Lunel-Viel (Hérault). Bull. Mus. Anthrop. préhist. Monaco 28, 7-62, Monaco.

1995: Le bison (Bovidae, Artiodactyla) du Pléistocène moyen ancien de Durfort (Gard, France). Bull. Mus. natn. Hist. nat., 4e sér., 16, C, 2-4, 349-381, Paris.

Brüning, H. 1970: Zur Klima-Stratigraphie der pleistozänen Mosbacher Sande bei Wiesbaden (Hessen). Mainzer Naturwiss. Archiv 9, 204-246, Mainz.

1978: Zur Untergliederung der Mosbacher Terrassenabfolge und zum klimatischen Stellenwert der Mosbacher Tierwelt im Rahmen des Cromer-Komplexes. Mainzer Naturwiss. Archiv 16, 143-190, Mainz.

Bukhsianidze, M. 2005: The Fossil Bovidae of Dmanisi. PhD Thesis, 192 p., Universitá degli Studi di Ferrara, Dipartimento delle Risorse Naturali e Culturali, Ferrara.

Caloi, L., Kotsakis, T., Palombo, M. R. & Petronio, C. 1996: The Pleistocene dwarf elephants of Mediterranean islands. In: Shoshani, J. & Tassy, P. (eds.): The Proboscidea, Evolution and Palaeoecology of Elephants and Their Relatives, 234-239, Oxford Univ. Press, Oxford etc.

Caloi, L. & Palombo, M. R. 1980: *Megaceros savini* e *Megaceros* cf. *verticornis* (Cervidae) del Pleistocene medio-inferiore di Ponte Galeria (Roma). Geol. Rom. 19, 121-130, Roma.

1986: Le mammalofaune plio-pleistoceniche dell'area laziale: problemi biostratigrafici ed implicazioni paleoclimatiche. Mem. Soc. Geol. Ital. 35, 99-126, Roma.

1995: Late Early Pleistocene Mammal Faunas of Italy: Biochronological Problems. Il Quaternario 8, 2, 391-402, Verona.

Campillo, D., Garcia Olivares, E., Malgosa, A., Martinez, F. & Martinez, B. (eds.) 1992: Proyecto Orce-Cueva Victoria (1988-1992). Presencia humana en el Pleistoceno inferior de Granada y Murcia. 503 p., Museo de Prehistoria José Gibert, Orce.

Campy, M. (dir.) 1983: L´Aven de Vergranne (Doubs). Un site du pléistocène moyen à Hominidé. Ann. Sci. Univ. Besançon, 4ème Sér., 5, 1-137, Besançon.

Cande, S. C. & Kent, D. V. 1995: Revised calibration of the geomagnetic polarity timescale for the Late Cretaceous and Cenozoic. J. Geophys. Res. 100, B 4, 6093-6095, Los Angeles.

Carbonell, E., Bermúdez de Castro, J. M., Arsuaga, J. L., Diez, J. C., Rosas, A., Cuenca-Bescós, G., Sala, R., Mosquera, M. & Rodrigues, X. P. 1995: Lower Pleistocene Hominids and Artifacts from Atapuerca-TD6 (Spain). Science 269, 826-830, Washington.

Cerdeño, E. 1993: Remarks on the Spanish Plio-Pleistocene *Stephanorhinus etruscus* (Rhinocerotidae). C. R. Acad. Sci. Paris 317, Sér. II, 1363-1367, Paris.

Chadwick, D. H. 1991: Elephants. Out of time, out of space. National Geographic 179, 5, 3-49, Washington.

Child, G. 1968: Behavior of large mammals during the formation of Lake Kariba. 123 p., Trustees of the National Museum of Rhodesia, Salisbury, Bulawayo.

Crégut, E. 1980a: *Equus mosbachensis tautavelensis* nov. subsp., un nouvel Equidé (Mammalia, Perissodactyla) du gisement Pléistocène moyen anté-rissien de la Caune de l'Arago (Tautavel, Pyrénées Orientales, France). Géobios 13, 1, 121-127, Lyon.

1980b: La faune de mammifères du gisement Pléistocène moyen anté-rissien de la Caune de l'Arago (Tautavel, Pyrénées-Orientales, France). C. R. Acad. Sci. Paris, Sér. D, 290, 751-754, Paris.

Crégut-Bonnoure, E. 1981: Données de la faune de grands mammiféres pour la biostratigraphie et l'environnement du gisement Pléistocène moyen de la Caune de l'Arago à Tautavel. In: Lumley, H. de & Labeyrie, J. (dir.): Colloque International du C. N. R. S., Tautavel, Lundi 22 - Dimanche 28 Juin 1981, Datations absolues et analyses isotopiques en Préhistoire, Méthodes et limites, Datation du remplissage de la Caune de l'Arago à Tautavel, 223-242, Tautavel.

1987: Caractères distinctifs du carpe et du tarse d'*Hemitragus* et de *Capra* (Mammalia, Bovidae, Caprinae). Bull. Mus. Anthrop. préhist. Monaco 30, 43-94, Monaco.

1999: Les petits Bovidae de Venta Micena (Andalousie) et de Cueva Victoria (Murcia). In: Gibert, J., Sánchez, F., Gibert, L. & Ribot, F. (eds.): Los Hominidos y su entorno en el Pleistoceno inferior y medio de Eurasia, Actas del Congreso Internacional de Paleontologia Humana, Orce 1995, 191-228, Museo Prehistoria y Paleontologia "J. Gibert", Orce.

Crégut, E. & Guérin, C. 1979: Première découverte en Europe sud-occidentale de *Praeovibos priscus* (Mammalia, Artiodactyla, Ovibovinae) dans le gisement Pléistocène moyen anté-rissien de la Caune de l'Arago (Tautavel, Pyrénées-Orientales, France). Géobios 12, 3, 459-465, Lyon.

Cremaschi, M., d'Henry, G., Peretto, C. & Sala, B. 1985: Isernia La Pineta. In: Homo, Journey to the Origins of Man`s History. Four Million Years of Evidence, 90-94, Cataloghi Marsilio, Venezia.

Croitor, R. 1997: Plio-Pleistocene deer from the Republic of Moldova. Their paleogeographic and stratigraphic significance. Author's abstract of Ph. D. thesis, 22 p., Inst. Zool. Acad. Sci. Moldova, Kshinau.

Croitor, R. & Kostopoulos, D. S. 2004: On the systematic position of the large-sized deer from Apollonia, Early Pleistocene, Greece. Paläont. Z. 78, 1, 137-159, Stuttgart.

Cuenca Bescós, G., Garcia, N. & Made, J. v. d. 2004: Fossil mammals of the Lower to Middle Pleistocene site of Trinchera Dolina, Atapuerca (Burgos, Spain). In: Baquedano, E. & Rubio Jara, S. (eds.): Zona Arqueológica, Miscelánea en homenaje a Emiliano Aguirre, Volumen II, Paleontologia, 140-149, Alcalá de Henares, Museo Arqueológico Regional, Madrid.

Czyzewska, T. 1989: Parzystokopytne – Artiodacty-la. In: Kowalski, K. (ed.): Historia i ewolucja ladowej fauny Polski, History and Evolution of the terrestrial fauna of Poland, Folia Quaternaria 59-60, 209-217, Kraków.

David, A. I. 1977: Sopostavlenie teriofauny antropogena Moldavii s sootvetsvuûŝimi faunami Evrazii. In: David, A. I., Negadaev-Nikonov, K. N., Hubka, A. N. & Azko, I. A. (eds.): Faunističeskie kompleksy i flora kajnozoâ Pričernomor'â, 3-23, Stiinza, Kišinev (in Russian).

1980: Malaeštskoe mestonahoždenie rannean-tropogenovoj fauny. Izv. Akad. Nauk Moldavskoj SSR, Ser. biol. him. nauk 3, 55-60, Kišinev (in Russian).

David, A. I. & Svistun, V. I. 1981: Ostatki bisonov iz verhnepliocenovyh i nižneplejstocenovyh otloženii Moldavii i ûga Ukrainy. In: Negadaev-Nikonov, K. N., David, A. I., Roska, V. H., Hubka, A. N. & Volosina, M. I. 1981: Biostratigrafiâ antropogena i neogena ûgo-zapada SSSR, 3-15, Stiinza, Kišinev (in Russian).

David, A. I., Tatarinov, K. A. & Svistun, V. I. 1990: Hiŝnye, hobotnye i kopytnye rannego plejsto-cena ûgo-zapada SSSR. 130 p., Stiinza, Kišinev (in Russian).

Daxner, G. 1968: Die Wildziegen (Bovidae, Mammalia) aus der altpleistozänen Karstspalte von Hundsheim in Niederösterreich. Ber. deutsch. Ges. geol. Wiss., A, Geol. Paläont., 13, 3, 305-334, Berlin.

Debenath, A. & Tournepiche, J.-F. 1993: Préhistoire de la Charente. Les temps anté-historiques en Angoumois à travers les collections du Musée. 79 p., Germa, Angoulême.

Deckert, K. & Karrer, C. 1965: Die Fischreste des Frühpleistozäns von Voigtstedt in Thüringen. Paläont. Abh., A, Paläozool., II, 2/3, 299-333 +pls. III-V, Berlin.

Delagnes, A., Tournepiche, J.-F., Armand, D., Desclaux, E., Diot, M.-F., Ferrier, C., Le Fillâtre, V. & Vandermeersch, B. 1999: Le gisement Pléistocène moyen et supérieur d'Artenac (Saint-Mary, Charente): premier bilan interdisci-plinaire. Bull. Soc. Préhist. Franc., 96, 4, 469-496, Paris.

Delson, E. 1980: Fossil Macaques, Phyletic Relationships and a Scenario of Deployment. In: Lindburg, D. G. (ed.): The Macaques: Studies in Ecology, Behavior and Evolution, 10-30, Van Nostrand Reinhold Comp., New York etc.

deMenocal, P. B. & Bloemendal, J. 1995. Plio-Pleistocene Climatic Variability in Subtropical Africa and the Paleoenvironment of Hominid Evolution: A Combined Data-Model Approach. In: Vrba, E. S., Denton, G. H., Partridge, T. C. & Burckle, L. H. (eds.): Paleoclimate and Evolu-tion, with Emphasis on Human Origins, 262-288, Yale Univ. Press, New Haven, London.

Dietrich, W. O. 1965: Fossile Elefantenzähne von Voigtstedt in Thüringen. Paläont. Abh., A, Paläozool., II, 2/3, 521-536 + pls. XXXII-XXXV, Berlin.

Drees, M. 1999: *Bison menneri* from the North Sea? Cranium 16, 2, 69-70, Leiden.

Dubrovo, I. A. 1955: Novye dannye o stroenii i rasprostranenii drevnego slona (*Hesperoloxo-don*). Dokl. AN SSSR 101, 4, 759- 762, Moskva (in Russian).

1960: Drevnie slony SSSR. Vypusk 1. Trudy

Paleont. Inst. 85, 1-78 + 3 pls., Moskva (in Russian).

1963: Novye dannye o tamanskom faunističeskom komplexe pozvonočnyh. Bûl. Moskovsk. Obs. ispyt. prir., otd. geol. 6, 94-99, Moskva (in Russian).

1964: Slony roda *Archidiskodon* na territorii SSSR. Paleont. Žurn. 3, 82-94, Moskva (in Russian).

1975: *Mammuthus trogontherii trogontherii* (Pohlig) aus dem Pleistozän von Tiraspol. Quartärpaläontologie 1, 125-155, Berlin.

2001: Remains of Elephantidae from the Lower Pleistocene site of Untermassfeld. In: Kahlke, R.-D. (ed.): Das Pleistozän von Untermaßfeld bei Meiningen (Thüringen), Part 2, Monogr. Röm.-Germ. Zentralmus. Mainz 40, 2, 589-605 + pls. 95-191, Habelt, Bonn.

Dubrovo, I. A. & Alekseev, M. N. 1964: K stratigrafii četvertičnyh otloženij Priazov'â. Bûll. Kom. Izuč. Četv. Per. 29, 35-43, Moskva (in Russian).

Dubrovo, I. A. & Kapelist, K. V. 1977: Katalog mestonahoždenij tretičnyh pozvonočnyh USSR. 159 p., Nauka, Moskva (in Russian).

Dubrovo, I. A. & Nigarov, A. N. 1990: Plio-Pleistocene Fossil Vertebrate Localities of South-Western Turkmenia, U.S.S.R. Quartärpaläontologie 8, 35-45, Berlin.

Eisenmann, V. 1999: *Equus granatensis* of Venta Micena and evidence for primitive non-stenonid horses in the Lower Pleistocene. In: Gibert, J., Sánchez, F., Gibert, L. & Ribot, F. (eds.): Los Hominidos y su entorno en el Pleistoceno inferior y medio de Eurasia, Actas del Congreso Internacional de Paleontologia Humana, Orce 1995, 171-189, Museo Prehistoria y Paleontologia "J. Gibert", Orce.

Eisenmann, V., Crégut-Bonnoure, E. & Moigne, A.-M. 1985: *Equus mosbachensis* et les grands chevaux de la Caune de l'Arago et de Lunel-Viel: crâniologie comparée. Bull. Mus. natn. Hist. nat., 4è sér., 7, C, 2, 157-173, Paris.

Ellenberg, J. 1968: Die geologisch-geomorphologische Entwicklung des südwest-thüringischen Werragebietes im Pliozän und Quartär. PhD Thesis, 188 p., Friedrich-Schiller-Universität Jena, Jena.

1975: Schwermineralverteilung, Sedimentation und Stratigraphie im Quartär der mittleren Werra. Z. geol. Wiss. 3, 11, 1389-1404, Berlin.

1995: Neotektonik. In: Seidel, G. (ed.): Geologie von Thüringen, 418-422, Schweizerbart, Stuttgart.

2002: Der thüringische Zersatzgrobschotter des Unterpleistozäns. In: Kahlke, R.-D. & Wunderlich, J. (eds.): Tertiär und Quartär in Thüringen, Beitr. Geol. Thüringen (N. F.) 9, 233-246, Jena.

Ellenberg, J. & Kahlke, R.-D. 1997: Die quartärgeologische Entwicklung des mittleren Werratals und der Bau der unterpleistozänen Komplexfundstelle Untermaßfeld. In: Kahlke, R.-D. (ed.): Das Pleistozän von Untermaßfeld bei Meiningen (Thüringen), Part 1, Monogr. Röm.-Germ. Zentralmus. Mainz 40, 1, 29-62 + pls. 1-18, Habelt, Bonn.

Ellenberg, J., Kahlke, R.-D. & Maul, L. 1997: Litho- und Biostratigraphie der thüringischen Quartär-Abfolge. In: Lützner, H. & Seidel, G. (eds.): Regionale Geologie von Mitteleuropa, Exkursionsführer, 149. Hauptversammlung der Deutschen Geologischen Gesellschaft, Schriftenr. Deutsch. Geol. Ges. 3, 35-57, Hannover.

Eltringham, S. K. 1999: The Hippos. Natural History and Conservation. 184 p., T. & A. D. Poyser Nat. Hist., London.

Essen, H. v. 2003: Tooth morphology of *Mammuthus meridionalis* from the southern bight of the North Sea and from several localities in the Netherlands. In: Reumer, J. W. F., Vos, J. de & Mol, D. (eds.): Advances in Mammoth Research, Proceedings of the Second International Mammoth Conference, Rotterdam, May 16-20 1999, Deinsea 9, 453-511, Rotterdam.

Falguères, C., Bahain, J.-J., Yokoyama, Y., Arsuaga, J. L., Bermúdez de Castro, J. M., Carbonell, E., Bischoff, J. L. & Dolo, J.-M. 1999: Earliest humans in Europe: the age of TD6 Gran Dolina, Atapuerca, Spain. J. Human Evol. 37, 3/4, 343-352, London etc.

Faure, M. & Guérin, C. 1984: *Sus strozii* et *Sus scrofa*, deux mammifères artiodactyles, marqueurs des paléoenvironnements. Palaeogeogr., Palaeoclim., Palaeoecol. 48, 215-228, Amsterdam etc.

Fejfar, O. 1961: Review of Quaternary vertebrata in Czechoslovakia. Inst. Geol. 34, INQUA

Warszawa 1961, 109-118 + pls. I-V, Warszawa.

1969: Human remains from the Early Pleistocene in Czechoslovakia. Current Anthropol. 10, 170-173, Chicago.

1993 (mit einem Beitrag von Wighart v. Koenigswald): Die Fauna aus den limnischen Ablagerungen von Přezletice bei Prag und ihre biochronologische Aussage. Jb. Röm.-Germ. Zentralmus. Mainz 40, 103-113 + pls. 4-8, Habelt, Bonn.

Fejfar, O. & Heinrich, W.-D. 1981: Zur biostratigraphischen Untergliederung des kontinentalen Quartärs anhand von Arvicoliden (Rodentia, Mammalia). Eclogae geol. Helv. 74, 3, 997-1006, Basel.

1990: Muroid Rodent Biochronology of the Neogene and Quaternary in Europe. In: Lindsay, E. H., Fahlbusch, V. & Mein, P. (eds.): European Neogene Mammal Chronology, NATO ASI Ser. A: Life Sci. 180, 91-117, Plenum, New York, London.

Fejfar, O. & Horáček, I. 1983: Zur Entwicklung der Kleinsäugerfaunen im Villányium und Alt-Biharium auf dem Gebiet der ČSSR. In: Heinrich, W.-D. (ed.): Wirbeltier-Evolution und Faunenwandel im Känozoikum, Schriftenr. geol. Wiss. 19/20, 111-207, Berlin.

Ferretti, M. 1999: *Mammuthus meridionalis* (Mammalia, Proboscidea, Elephantidae) from the "Sabbie Gialle" of Oriolo (Cava La Salita, Faenza, Northern Italy) and other European late populations of southern mammoth. Eclogae geol. Helv. 92, 503-515, Basel.

Fischer, K. 1985: Säugetierfunde aus dem Quartär des Kreises Artern. Beiträge zur Kyffhäuserlandschaft, Veröff. Kreisheimatmus. Bad Frankenhausen 9, 69-96, Bad Frankenhausen.

Fladerer, F. 1987: *Macaca* (Cercopithecidae, Primates) im Altpleistozän von Deutsch-Altenburg, Niederösterreich. Beitr. Paläont. Österr. 13, 1-24, Wien.

1991: Der erste Fund von *Macaca* (Cercopithecidae, Primates) im Jungpleistozän von Mitteleuropa. Z. Säugetierkd. 56, 272-283, Hamburg, Berlin.

Flerov, K. K. 1979: Sistematika i evolûziâ. In: Sokolov, V. E. (ed.): Zubr. Morfologiâ, sistematika, evolûziâ, ekologiâ, 9-127, Nauka, Moskva (in Russian).

Forstén, A. 1986: A Review of the Süssenborn Horses and the Origin of *Equus hydruntinus* Regalia. Quartärpaläontologie 6, 43-52, Berlin.

1988: Middle Pleistocene Replacement of Stenonid Horses by Caballoid Horses - Ecological Implications. Palaeogeogr., Palaeoclim., Palaeoecol. 65, 23-33, Amsterdam etc.

1990: Old World "Asses". Quartärpaläontologie 8, 71-78, Berlin.

1998: Equus species as stratigraphic markers. Reality or wishful thinking? Quaternary Sci. Rev. 17, 1097-1100, Oxford.

1999: A review of *Equus stenonis* Cocchi (Perissodactyla, Equidae) and related forms. Quaternary Sci. Rev. 18, 1373-1408, Oxford.

Forsten, A. & Sharapov, S. 2000: Fossil equids (Mammalia, Equidae) from the Neogene and Pleistocene of Tadzhikistan. Geodiversitas 22, 2, 293-314, Paris.

Fortelius, M., Mazza, P. & Sala, B. 1993: *Stephanorhinus* (Mammalia: Rhinocerotidae) of the western European Pleistocene, with a revision of *S. etruscus* (Falconer, 1868). Palaeontograph. Ital. 80, 63-155, Pisa.

Fosse, P. 1995: Le rôle de l'hyène dans la formation des assemblages osseux: 150 ans de controverses. L'apport des anciens textes de préhistoire et de paléontologie du Quaternaire aux études taphonomiques actuelles. Paléo 7, 49-84, Les Eyzies.

1996: La grotte n° 1 de Lunel-Viel (Hérault, France): Repaire d'hyènes du Pléistocène moyen. Étude taphonomique du matériel osseux. Paléo 8, 47-81, Les Eyzies.

Frank, C. & Rabeder, G. 1997a: Deutsch-Altenburg 2-4-16-30. In: Döppes, D. & Rabeder, G. (eds.): Pliozäne und pleistozäne Faunen Österreichs. Ein Katalog der wichtigsten Fossilfundstellen und ihrer Faunen, Mitt. Komm. Quartärforsch. Österr. Akad. Wiss. 10, 241-250, Wien.

1997b: Hundsheim. In: Döppes, D. & Rabeder, G. (eds.): Pliozäne und pleistozäne Faunen Österreichs. Ein Katalog der wichtigsten Fossilfundstellen und ihrer Faunen, Mitt. Komm. Quartärforsch. Österr. Akad. Wiss. 10, 270-274, Wien.

Franzen, J. L. 1973: Ein Primate aus den altpleistozänen Schneckenmergeln von Hohensülzen (Rheinhessen). Senckenbergiana lethaea 54, 2/4, 345-358, Frankfurt a. M.

1999a: Die große Flut – der Rheinhessensee. Natur u. Museum 129, 7, 201-212, Frankfurt a. M.

1999b: The late Early Pleistocene teeth and bone accumulation of Dorn-Dürkheim 3 (Germany, Rheinhessen): Natural or manmade? In: The role of early humans in the accumulation of European Lower and Middle Palaeolithic bone assemblages, Monogr. Röm.-Germ. Zentralmus. Mainz 42, 41-56, Habelt, Bonn.

Franzen, J. L., Gliozzi, E., Jellinek, T., Scholger, R. & Weidenfeller, M. 2000: Die spätaltpleistozäne Fossillagerstätte Dorn-Dürkheim 3 und ihre Bedeutung für die Rekonstruktion der Entwicklung des rheinischen Flußsystems. Senckenbergiana lethaea 80, 1, 305-353, Frankfurt a. M.

Gabounia, L. & Vekua, A. K. 1995: La mandibule de l'homme fossile du Villafranchien supérieur de Dmanissi (Géorgie orientale). L'Anthropologie 99, 1, 29-41, Paris.

Gabunia, L., Jöris, O., Justus, A., Lordkipanidze, D., Muschelisvili., A., Nioradze, M., Swisher III, C. C. & Vekua, A. K., unter Mitarbeit von Bosinski, G., Ferring, R. C., Majsuradze, G. M. & Tvalcrelidze, M. 1999a: Neue Hominidenfunde des altpaläolithischen Fundplatzes Dmanisi (Georgien, Kaukasus) im Kontext aktueller Grabungsergebnisse. Archäol. Korrespondenzbl. 29, 4, 451-488, Mainz.

Gabunia, L. K., Jöris, O., Justus, A., Lordkipanidze, D., Nioradze, M. & Vekua, A. 1999-2000: *Homo erectus* aux portes de l'Europe. Découvertes d'hominidés de 1,75 million d'années à Dmanisi (Géorgie, Caucase). L'Archéologue 45, 46-50, Paris.

Gabunia, L., Justus, A., Vekua, A. & Lordkipanidze, D. 1999b: Ein neu entdeckter Hominiden-Rest vom Fundplatz Dmanisi (Ostgeorgien). Archäol. Korrespondenzbl. 29, 3, 299-305, Mainz.

Gabunia, L. & Vekua, A. 1989: Antropogenovye lošadi Gruzii. 140 p., Mezniereba, Tbilisi (in Russian).

1990: L'évolution du paléoenvironnement au cours de l'anthropogène en Géorgie (Transcaucasie). L'Anthropologie 94, 4, 643-650, Paris.

1993: Dmanisskij iskopaemyj čelovek i soputstvuûsaâ emu fauna pozvonočnyh. 72 p., Mezniereba, Tbilisi (in Russian).

Gabunia, L., Vekua, A., Lordkipanidze, D., Swisher III, C. C., Ferring, R., Justus, A., Nioradze, M., Tvalchrelidze, M., Antón, S. C., Bosinski, G., Jöris, O., Lumley, M.-A. de, Majsuradze, G. & Mouskhelishvili, A. 2000: Earliest Pleistocene Hominid Cranial Remains from Dmanisi, Republic of Georgia: Taxonomy, Geological Setting, and Age. Science 288, 1019-1025, Washington.

Gabunia, M. K., Vekua, A. K. & Tvalcrelidze, M. 1994: Der altpaläolithische Fundplatz Achalkalaki am Amiranis-Berg im Dzavacheti-Gebiet (Süd-Georgien). Archäol. Korrespondenzbl. 24, 237-240, Mainz.

Gagnepain, J., Hedley, I., Bahain, J.-J., Falguères, C., Laurent, M., Peretto, C., Wagner, J.-J. & Yokoyama, Y. 1996: Synthèse des données radiochronologiques et paléomagnétiques du site de Ca'Belvedere di Monte Poggiolo (Romagna, Italie) et de son environnement géologique. In: XIII Int. Congr. Prehist. Protohist. Sci. Forli, Italia, 8/14 Sept. 1996, Abstracts 2, The workshops and the posters of the XIII International Congress of Prehistoric and Protohistoric Sciences, 129-130, Abaco s. r. l., Forli.

Galobart, À. & Maroto, J. (eds.) 2003: Los yacimientos paleontólogicos de Incarcal (Crespià). Paleont. i Evol. 34, 1-298, Barcelona.

Garcia Garcia, N. 2003: Osos y otros carnívoros de la Sierra de Atapuerca. 575 p., Fundación Oso de Asturias, Oviedo.

Garcia, N. 2004: New results on the remains of Ursidae from Untermaßfeld: comparisons with *Ursus dolinensis* from Atapuerca and other Early and Middle Pleistocene sites. In: Maul, L. C. & Kahlke, R.-D. (eds.): Late Neogene and Quaternary biodiversity and evolution: Regional developments and interregional correlations, Conference Volume, 18th International Senckenberg Conference, VI International Palaeontological Colloquium in Weimar, Weimar (Germany), 25th - 30th April, 2004, Terra Nostra, Schriften der Alfred-Wegener-Stiftung 2004/2, 112-113, Berlin.

Garcia, N. & Arsuaga, J. L. 1999: Carnivores from

the Early Pleistocene hominid-bearing Trinchera Dolina 6 (Sierra de Atapuerca, Spain). J. Human Evol. 37, 3/4, 415-430, London etc.

2001a: Les carnivores (Mammalia) des sites du Pléistocène ancien et moyen d'Atapuerca (Espagne). L'Anthropologie 105, 83-93, Paris.

2001b: *Ursus dolinensis*: a new species of Early Pleistocene ursid from Trinchera Dolina, Atapuerca (Spain). C. R. Acad. Sci. Paris, Sci. de la Terre et des planètes 332, 717-725, Paris.

Garutt, V. E. 1954: Ûznyj slon *Archidiskodon meridionalis* (Nesti) iz pliozena severnogo poberež'â Azovskogo Morâ. Bûll. Kom. Izuč. Četv. Per. 10, 2, 1-76, Moskva, Leningrad (in Russian).

Garutt, V. E. & Bajguševa, V. S. 1981: *Archidiskodon gromovi* Garutt et Alexeeva – der älteste Elefant der Mammutlinie in Eurasien. Quartärpaläontologie 4, 19-25, Berlin.

Garutt, V. E. & Safronov, I. N. 1965: Nahodka skeleta ûznogo slona *Archidiskodon meridionalis* (Nesti) bliz g. Georgievska (Severnyj Kavkaz). Bûll. Kom. Izuč. Četv. Per. 30, 79-88, Moskva (in Russian).

Gaudant, J. 1979: L`ichthyofaune tiglienne de Tegelen (Pays-Bas): signification paléoécologique et paléoclimatique. Scripta Geologica 50, 1-16, Leiden.

Gentili, S., Abbazzi, L., Masini, F., Ambrosetti, P., Argenti, P. & Torre, D. 1996: Voles from the Early Pleistocene of Pietrafitta (central Italy, Perugia). In: Nadachowski, A. & Werdelin, L. (eds.): Neogene and Quaternary Mammals of the Palaearctic, Papers in Mammal Palaeontology honoring Kazimierz Kowalski, Acta zool. cracov. 39, 1, 185-199, Kraków.

Geraads, D. 1983: Artiodactyles (Mammalia) du Pléistocène de Vergranne (Doubs). In: Campy, M. (dir.): L´Aven de Vergranne (Doubs), Un site du pléistocène moyen à Hominidé, Ann. Sci. Univ. Besançon, 4ème Sér., 5, 69-81, Besançon.

1986: Les ruminants du Pléistocène d'Oubeidiyeh (Israel). In: Tchernov, E. & Guérin, C. (eds.): Les mammifères du Pléistocène inférieur de la vallée du Jourdain à Oubéidiyeh, Mém. trav. Centre Rech. franç. Jérusalem 5, 143-181, Paris.

1990: Contribution des cervidés à la chronologie des débuts de l'occupation humaine en Europe occidentale. Quaternaire 1, 3-4, 167-174, Paris.

Gibbard, P. L., West, R. G., Zagwijn, W. H., Balson, P. S., Burger, A. W., Funnell, B. M., Jeffery, D. H., Jong, J. de, Kolfschoten, T. v., Lister, A. M., Meijer, T., Norton, P. E. P., Preece, R. C., Rose, J., Stuart, A. J., Whiteman, C. A. & Zalasiewicz, J. A. 1991: Early and early Middle Pleistocene correlations in the Southern North Sea Basin. Quaternary Sci. Rev. 10, 23-52, Oxford.

Gibbard, P. L., Zalasiewicz, J. A. & Mathers, S. J. 1998: Stratigraphy of the marine Plio-Pleistocene crag deposits of East Anglia. In: Kolfschoten, T. v. & Gibbard, P. L. (eds.): The Dawn of the Quaternary, Proceedings of the SEQS-EuroMam symposium 1996, Meded. Nederl. Inst. Toegepaste Geowetenschappen TNO 60, 239-262, Haarlem.

Giuli, C. de 1987: Late Villafranchian faunas of Italy: the Selvella Local Fauna in the southern Chiana Valley – Umbria. Palaeontograph. Ital. 74, 11-50, Pisa.

Giuli, C. de & Masini, F. 1987: Late Villafranchian faunas in Italy: the Casa Frata Local Fauna (Upper Valdarno, Tuscany). Palaeontograph. Ital. 74, 1-9, Pisa.

Giuli, C. de, Masini, F. & Torre, D. 1987: The Latest Villafranchian Faunas in Italy: the Pirro Nord Fauna (Apricena, Gargano). Palaeontograph. Ital. 74, 51-62, Pisa.

Goddard, J. 1970: Age criteria and vital statistics of a Black Rhinoceros population. E. Afr. Wildl. J. 8, 105-121, Oxford.

Grasselt, T. 1994: Siedlungs- und Grabfunde von der „Strick" bei Henfstädt, Lkr. Hildburghausen. In: Dušek, S. (ed.): Führer zu archäologischen Denkmälern in Deutschland, Vol. 28: Südliches Thüringen, 170-173, Theiss, Stuttgart.

Gromova, V. I. & Dubrovo, I. A. 1975: Fossile Pferde aus dem Faunenkomplex von Tiraspol. Quartärpaläontologie 1, 107-124, Berlin.

Guenther, E. W. 1989: Die Bestimmung der Art und der zeitlichen Eingliederung pleistozäner Elefanten nach ihren Gebissen. Zur Untersuchung einer eiszeitlichen Elefantenfauna von Randersacker bei Würzburg. Quartär 39/40, 7-65, Bonn.

Guenther, E. W. & Mai, H. 1977: Die pleistozänen Schichten von Jockgrim in der Rheinpfalz. Schr. Naturw. Ver. Schlesw.-Holst. 47, 5-24, Kiel.

Guérin, C. 1990: Biozones or mammal units? Methods and limits in Biochronology. In: Lindsay, E. H., Fahlbusch, V. & Mein, P. (eds.): European Neogene Mammal Chronology, NATO ASI Ser. A, Life Sci. 180, 119-130, Plenum, New York, London.

Guérin, C. & Barychnikov, G. F. 1987: Le rhinocéros acheuléen de la grotte de Koudaro I (Géorgie, URSS) et le problème des espèces relictes du Pléistocène du Caucase. Géobios 20, 3, 389-396, Lyon.

Guérin, C., Bar-Yosef, O., Debard, E., Faure, M., Shea, J. & Tchernov, E. 1996: Mission archéologique et paléontologique dans le Pléistocène ancien d'Oubéidiyeh (Israël): résultats 1992-1994. C. R. Acad. Sci. Paris 322, Sér. II a, 709-712, Paris.

Guérin, C., Dewolf, Y. & Lautridou, J.-P. 2003: Révision d'un site paléontologique célèbre: Saint-Prest (Chartres, France). Géobios 36, 1, 55-82, Villeurbanne.

Guérin, C., Eisenmann, V. & Faure, M. 1993: Les grands mammifères du gisement pléistocène moyen de Latamné (vallée de l'Oronte, Syrie). In: Sanlaville, P., Besançon J., Copeland, L. & Muhesen, S. (dir.): Le Paléolithique de la vallée moyenne de l'Oronte (Syrie), Peuplement et environnement, BAR Internat. Ser. 587, 169-178, Oxford.

Guérin, C. & Faure, M. 1997: The Wild Boar (*Sus scrofa priscus*) from the Post-Villafranchian Lower Pleistocene of Untermaßfeld. In: Kahlke, R.-D. (ed.): Das Pleistozän von Untermaßfeld bei Meiningen (Thüringen), Part 1, Monogr. Röm.-Germ. Zentralmus. Mainz 40, 1, 375-383 + pls. 63-67, Habelt, Bonn.

Gülec, E., Howell, F. C. & White, T. D. 1999: Dursunlu – A new Lower Pleistocene Faunal and Artifact-bearing Locality in Southern Anatolia. In: Ullrich, H. (ed.): Hominid Evolution, Lifestyles and Survival Strategies, 349-364, Edition Archaea, Gelsenkirchen / Schwelm.

Haltenorth, T. & Diller, H. 1977: Säugetiere Afrikas und Madagaskars. 403 p., BLV Verlagsges., München, Bern.

Haynes, G. 1988: Mass Death and Serial Predation: Comparative Taphonomic Studies of Modern Large Mammal Death Sites. J. Archaeol. Sci. 15, 219-235, London etc.

Hedberg, H. D. (ed.) 1976: International Stratigraphic Guide. 200 p., J. Wiley & Sons, New York.

Heller, F. 1936: Eine Forest Bed-Fauna aus der Schwäbischen Alb. Sitzungsber. Heidelberger Akad. Wiss., Math.-nat. Kl., 2, 1-29, Heidelberg.

1958: Eine neue altquartäre Wirbeltierfauna von Erpfingen (Schwäbische Alb). N. Jb. Geol. Paläont. Abh. 107, 1, 1-102, Stuttgart.

Hemmer, H. 1971: Zur Charakterisierung und stratigraphischen Bedeutung von *Panthera gombaszoegensis* (Kretzoi, 1938). N. Jb. Geol. Paläont. Mh., 12, 701-711, Stuttgart.

1972: Zur systematischen Stellung von „*Jansofelis vaufreyi*" Bonifay, 1971, und „*Felis lunellensis*" Bonifay, 1971, aus dem Pleistozän Südfrankreichs (Carnivora, Felidae). N. Jb. Geol. Paläont. Mh., 4, 215-223, Stuttgart.

1993a: *Felis silvestris* Schreber, 1777 – Wildkatze. In: Stubbe, M. & Krapp, F. (eds.): Handbuch der Säugetiere Europas, Vol. 5: Raubsäuger – Carnivora (Fissipedia), Part II: Mustelidae 2, Viverridae, Herpestidae, Felidae, 1076-1118, Aula-Verl., Wiesbaden.

1993b: *Felis (Lynx) lynx* Linnaeus, 1758 – Luchs, Nordluchs. In: Stubbe, M. & Krapp, F. (eds.): Handbuch der Säugetiere Europas, Vol. 5: Raubsäuger – Carnivora (Fissipedia), Part II: Mustelidae 2, Viverridae, Herpestidae, Felidae, 1119-1167, Aula-Verl., Wiesbaden.

2000: Out of Asia: A paleoecological scenario of man and his carnivorous competitors in the European Lower Pleistocene. In: Lordkipanidze, D., Otte, M. & Bar-Yosef, O. (eds.): Early Humans at the Gates of Europe, Proceedings of the first international symposium. Dmanisi, Tbilisi (Georgia), Septembre 1998, Et. Rech. Archéol. Univ. Liège 92, 99-106, Liège.

2001: Die Feliden aus dem Epivillafranchium von Untermaßfeld. In: Kahlke, R.-D. (ed.): Das Pleistozän von Untermaßfeld bei Meiningen (Thüringen), Part 3, Monogr. Röm.-Germ. Zentralmus. Mainz 40, 3, 699-782 + pls. 132-143, Habelt, Bonn.

2002: Pleistozäne Katzen Europas – eine

Übersicht. Cranium 20, 2, 6–22, Leiden.

Hemmer, H. & Kahlke, R.-D. 2005: Nachweis des Jaguars (*Panthera onca gombaszoegensis*) aus dem späten Unter- oder frühen Mittelpleistozän der Niederlande. Deinsea 11, 1-11, Rotterdam.

Hemmer, H., Kahlke, R.-D. & Keller, T. 2003: *Panthera onca gombaszoegensis* (Kretzoi, 1938) from the Early Middle Pleistocene Mosbach Sands (Wiesbaden, Hessen, Germany) – A contribution to the knowledge of the variability and history of the dispersal of the jaguar. N. Jb. Geol. Paläont. Abh. 229, 1, 31-60, Stuttgart.

Hemmer H., Kahlke, R.-D. & Vekua, A. K. 2001: The Jaguar - *Panthera onca gombaszoegensis* (Kretzoi, 1938) (Carnivora: Felidae) in the late Lower Pleistocene of Akhalkalaki (South Georgia; Transcaucasia) and its evolutionary and ecological significance. Géobios 34, 4, 475-486, Villeurbanne.

Hemmer, H., Kahlke, R.-D. & Vekua, A. K., 2004 : The Old World puma – *Puma pardoides* (Owen, 1846) (Carnivora: Felidae) – in the Lower Villafranchian (Upper Pliocene) of Kvabebi (East Georgia, Transcaucasia) and its evolutionary and biogeographical significance. N. Jb. Geol. Paläont. Abh. 233, 2, 197-231, Stuttgart.

Hemmer, H. & Schütt, G. 1969: Ein Unterkiefer von *Panthera gombaszoegensis* (Kretzoi, 1938) aus den Mosbacher Sanden. Mainzer Naturw. Arch. 8, 90-101, Mainz.

1970: Körpergröße und Extremitätenmaße ältest- und altpleistozäner europäischer Pantherkatzen (Genus *Panthera*). Mainzer Naturw. Arch. 9, 132-146, Mainz.

Heptner, V. G., Naumov, N. P., Jürgenson, P. B., Sludski, A. A., Cirkova, A. F. & Bannikov, A. G. 1974: Die Säugetiere der Sowjetunion. Vol. II: Seekühe und Raubtiere. 1006 p., Fischer, Jena.

Holec, P. 1985: Vertebratenfauna des Quartärs in der Slowakei. Acta Geol. Geogr. Univ. Comenianae, Geologica 39, 115-128, Bratislava.

Hooijer, D. A. 1959: Fossil mammals from Jisr Banat Yaqub, south of lake Huleh, Israel. Bull. Res. Counc. Israel, 8, G: Geo-Sci., 4, 177-199, Jerusalem.

1960: A *Stegodon* from Israel. Bull. Res. Counc. Israel, 9, G: Geo-Sci., 104-108, Jerusalem.

1961: Middle Pleistocene mammals from Latamne, Orontes valley, Syria. Ann. Archéol. Syrie 11, 117-132, Damaskus.

1966: Additional notes on the Pleistocene Mammalian fauna of the Orontes valley. Ann. Archéol. Syrie 15, 101-104, Damaskus.

Horáček, I. 1980: Včeláre 3 - nové naleziste staropleistocenni fauny Slovenského krasu. Slov. kras 18, 183-192, Brno.

1985: Survey of the fossil vertebrate localities Včeláre 1 - 7. Čas. min. geol. 30, 4, 353-366, Praha.

Horwitz, L. K. 1998: The influence of prey body size on patterns of bone distribution and representation in a striped hyaena den. In: Économie préhistorique: Les comportements de subsistance au Paléolithique, XVIIIe Recontres Internationales d'Archéologie et d'Histoire d'Antibes, 31-42, Éd. APDCA, Sophia Antipolis.

Howell, F. C., White, T. D., Gülec, E., Sarac, G. & Curtis, G. H. 1999: Dursunlu, Lower Pleistocene faunal and archeological locality, Konya basin, Anatolia (Turkey). In: Gibert, J., Sánchez, F., Gibert, L. & Ribot, F. (eds.): Los Hominidos y su entorno en el Pleistoceno inferior y medio de Eurasia, Actas del Congreso Internacional de Paleontologia Humana, Orce 1995, 459-468, Museo Prehistoria y Paleontologia "J. Gibert", Orce.

Jaarsveld, A. S. v. 1993: A comparative investigation of hyaenid and aardwolf life-histories, with notes on spotted hyaena mortality patterns. Trans. Royal Soc. South Africa 48, 2, 219-231, Cape Town.

Janis, C. M. 2003: Correlation of cranial and dental variables with body size in ungulates and macropodoids. In: Damuth, J. & MacFadden, B. J. (eds.): Body Size in Mammalian Paleobiology: Estimation and Biological Implications, 255-299, Cambridge University Press, Cambridge etc.

Jánossy, D. 1962: Az elsö fosszilis vizilovleletek hazank Pleisztocénjéböl. Der erste Nachweis von *Hippopotamus antiquus* Desmarest, 1822 im ungarischen Altpleistozän (Budapest). Különl. Állatt. Közlém. 49, 1, 63-74 + pls. I-II, Budapest.

1963: Die altpleistozäne Wirbeltierfauna von Kövesvárad bei Répáshuta (Bükk-Gebirge).

Ann. Hist.-Nat. Mus. Natn. Hungarici 55, Pars Min. Palaeont., 109-141, Budapest.

1970: Ein neuer Eomyide (Rodentia, Mammalia) aus dem Ältestpleistozän («Oberes Villafrankium», Villányium) des Osztramos (Nordostungarn). Ann. Hist.-Nat. Mus. Natn. Hungarici 62, Pars Min. Palaeont., 99-113, Budapest.

1972: Ein kleiner *Hystrix* aus dem Altpleistozän der Fundstelle Osztramos 8. (Nordungarn). Vertebr. Hung. 13, 163-182, Budapest.

1983: Lemming-remain from the Older Pleistocene of Southern Hungary (Villány, Somssich-hegy 2). Fragm. Min. Palaeont. 11, 55-60, Budapest.

1986: Pleistocene vertebrate faunas of Hungary. 208 p., Akad. Kiadó, Budapest.

1990: Vertebrate fauna of site II. In: Kretzoi, M. & Dobosi, V. T. (eds.): Vértesszölös. Man, Site and Culture, 187-229, Akad. Kiadó, Budapest.

1997: Die Vogelreste aus dem Unterpleistozän von Untermaßfeld. In: Kahlke, R.-D. (ed.): Das Pleistozän von Untermaßfeld bei Meiningen (Thüringen), Part 1, Monogr. Röm.-Germ. Zentralmus. Mainz 40, 1, 95-100 + pl. 21, Habelt, Bonn.

Kahlke, H.-D. 1961: Revision der Säugetierfaunen der klassischen deutschen Pleistozän-Fundstellen von Süßenborn, Mosbach und Taubach. Geologie 10, 4/5, 493-532, Berlin.

1963: *Ovibos* aus den Kiesen von Süßenborn. Ein Beitrag zur Systematik und Phylogenie der Ovibovini und zur Stratigraphie des Pleistozäns. Geologie, 12, 8, 942-972, Berlin.

1964: Early middle pleistocene (Mindel/Elster) *Praeovibos* and *Ovibos*. Soc. Sci. Fenn. Comm. Biol. 24, 5, 3-17, Helsinki.

(ed.) 1965: Das Pleistozän von Voigtstedt. Paläont. Abh., A, II, 2/3, 221-692 + pls. I-XL, Berlin.

(ed.) 1969: Das Pleistozän von Süßenborn. Paläont. Abh., A, III, 3/4, 367-788 + pls. I-LXVI, Berlin.

1975: The Macro-faunas of Continental Europe During the Middle Pleistocene: Stratigraphic Sequence and Problems of Intercorrelation. In: Butzer, K. W. & Isaac, G. L. (eds.): After the Australopithecines, 309-374, Mouton, The Hague.

(ed.) 1990a: Quartärpaläontologie 8. 1-281, Akademie-Verl., Berlin.

1990b: On the Evolution, Distribution and Taxonomy of Fossil Elk/Moose. Quartärpaläontologie 8, 83-106, Berlin.

1995a: Ein fossiler Elch-Fund aus dem Unterpleistozän von Untermaßfeld in Thüringen (Mitteldeutschland). Quartär 45/46, 227-235, Bonn.

1995b: Neue Cerviden-Reste der Stránská Skála bei Brno. In: Musil, R. (ed.): Stránská Skála Hill. Excavation of open-air sediments 1964-1972, Anthropos 26 (N. S. 18), 85-91, Brno.

1997: Die Cerviden-Reste aus dem Unterpleistozän von Untermaßfeld. In: Kahlke, R.-D. (ed.): Das Pleistozän von Untermaßfeld bei Meiningen (Thüringen), Part 1, Monogr. Röm.-Germ. Zentralmus. Mainz 40, 1, 181-275 + pl. 34-44, Habelt, Bonn.

2001a: Neufunde von Cerviden-Resten aus dem Unterpleistozän von Untermaßfeld. In: Kahlke, R.-D. (ed.): Das Pleistozän von Untermaßfeld bei Meiningen (Thüringen), Part 2, Monogr. Röm.-Germ. Zentralmus. Mainz 40, 2, 461-482 + pl. 72-76, Habelt, Bonn.

2001b: Die Rhinocerotiden-Reste aus dem Unterpleistozän von Untermaßfeld. In: Kahlke, R.-D. (ed.): Das Pleistozän von Untermaßfeld bei Meiningen (Thüringen), Part 2, Monogr. Röm.-Germ. Zentralmus. Mainz 40, 2, 501-555 + pl. 77-91, Habelt, Bonn.

Kahlke, R.-D. 1985a: Untersuchungen zur Incisivenreduktion an altpleistozänen *Hippopotamus*-Mandibeln von Untermaßfeld bei Meiningen (Bezirk Suhl, DDR). Biol. Rdsch. 23, 315-321, Jena.

1985b: On the Occurrence of *Hippopotamus* (Mammalia, Artiodactyla) in the Pleistocene of Achalkalaki (Gruzinian SSR, Soviet Union) and on the Distribution of the Genus in South-East Europe. Z. geol. Wiss. 15, 3, 407-414, Berlin.

1987: Die unterpleistozänen *Hippopotamus*-Reste von Untermaßfeld bei Meiningen (Bezirk Suhl, DDR) – Ein Beitrag zur Forschungs-, Entwicklungs- und Verbreitungsgeschichte fossiler Hippopotamiden in Europa. PhD Thesis,

167 p. + 21 suppl. + 102 tabs. + 27 figs. + 18 pls., Ernst-Moritz-Arndt-Universität Greifswald, Greifswald.

1989: Die unterpleistozänen *Hippopotamus*-Reste von Würzburg-Schalksberg. Quartär 39/40, 67-94, Bonn.

1990: Zum Stand der Erforschung fossiler Hippopotamiden (Mammalia, Artiodactyla) – Eine Übersicht. Quartärpaläontologie 8, 107-118, Berlin.

1992: Eine komplexe Großsäugerfauna aus der Forschungsgrabung Untermaßfeld (Südthüringen). In: DEUQUA ´92, Tagungsprogramm und Kurzfassungen, 59-61, Kiel.

1993a: Die Entstehungs-, Entwicklungs- und Verbreitungsgeschichte des oberpleistozänen *Mammuthus-Coelodonta*-Faunenkomplexes in Eurasien (Großsäuger). Habilitation Thesis, 285 p. + 23 maps, Christian-Albrechts-Univ. Kiel, Kiel.

1993b: Untermassfeld – A Complex Epivillafranchian Vertebrate Fauna from a Neogene Volcanic Area in Central Germany. In: Follieri, M., Girotti, O., Kotsakis, T., Taddeucci, A. & Turner, C. (eds.): Symposium Quaternary Stratigraphy in Volcanic Areas, Abstracts, INQUA SEQS, 39, Rome.

1994: Die Entstehungs-, Entwicklungs- und Verbreitungsgeschichte des oberpleistozänen *Mammuthus-Coelodonta*-Faunenkomplexes in Eurasien (Großsäuger). Abh. senckenberg. naturforsch. Ges. 546, 1-164, Frankfurt a. M. [= 1999: The History of the Origin, Evolution and Dispersal of the Late Pleistocene *Mammuthus-Coelodonta* Faunal Complex in Eurasia (Large Mammals). 219 p., Fenske, Rapid City.]

1995a: Untermassfeld near Meiningen – Early Pleistocene fauna. In: Schirmer, W. (ed.): Quaternary field trips in Central Europe. Volume 2. Field trips on special topics, 730-732, Pfeil, München.

1995b: Die Abfolge plio-/pleistozäner Säugetierfaunen in Thüringen (Mitteldeutschland). Cranium 12, 1, 5-18, Leiden.

1996: Geology, taphonomy and larger mammal fauna from the Lower Pleistocene Untermassfeld site (Central Germany). In: Kolfschoten, T. v. & Gibbard, P. (eds.): The Dawn of the Quaternary, Abstracts INQUA-SEQS, 16-21 June 1996, 49, Kerkrade.

(ed.) 1997a: Das Pleistozän von Untermaßfeld bei Meiningen (Thüringen), Part 1. Monogr. Röm.-Germ. Zentralmus. Mainz 40, 1, I-VIII + 1-418 + pls. 1-67, Habelt, Bonn.

1997b: Zur Entdeckungs- und Erforschungsgeschichte der unterpleistozänen Komplexfundstelle Untermaßfeld. In: Kahlke, R.-D. (ed.): Das Pleistozän von Untermaßfeld bei Meiningen (Thüringen), Part 1, Monogr. Röm.-Germ. Zentralmus. Mainz 40, 1, 1-28, Habelt, Bonn.

1997c: Die *Hippopotamus*-Reste aus dem Unterpleistozän von Untermaßfeld. In: Kahlke, R.-D. (ed.): Das Pleistozän von Untermaßfeld bei Meiningen (Thüringen), Part 1, Monogr. Röm.-Germ. Zentralmus. Mainz 40, 1, 277-374 + pls. 45-62, Habelt, Bonn.

1997d: Bisheriger Gesamtbefund zur Geologie, Paläozoologie, Taphonomie, Ökologie und Stratigraphie der unterpleistozänen Komplexfundstelle Untermaßfeld. In: Kahlke, R.-D. (ed.): Das Pleistozän von Untermaßfeld bei Meiningen (Thüringen), Part 1, Monogr. Röm.-Germ. Zentralmus. Mainz 40, 1, 385-418, Habelt, Bonn.

1999: Overview and first quantitative data on the taphonomy of the Lower Pleistocene fossil site of Untermaßfeld (Thüringen, Germany). In: The role of early humans in the accumulation of European Lower and Middle Palaeolithic bone assemblages. Monogr. Röm.-Germ. Zentralmus. Mainz 42, 7-19, Habelt, Bonn.

2000a: Untermassfeld – a reference fauna of the Eurasian Post-Villafranchian Lower Pleistocene. In: Les premiers habitants de l'Europe, Résumés des Communications, 52, Tautavel.

2000b: Beuterest: Vorderbein-Skelett eines «Etruskischen Nashorns» aus Südthüringen. Natur u. Museum 130, 6, 197-199, Frankfurt a. M.

2000c: The Early Pleistocene (Epivillafranchian) Faunal Site of Untermassfeld (Thuringia, Central Germany) - Synthesis of new results. In: Lordkipanidze, D., Otte, M. & Bar-Yosef, O. (eds.): Early Humans at the Gates of Europe, Proceedings of the first international symposium. Dmanisi, Tbilisi (Georgia), Septembre 1998, Et. Rech. Archéol. Univ. Liège 92, 123-138, Liège.

(ed.) 2001a: Das Pleistozän von Untermaßfeld

bei Meiningen (Thüringen), Part 2. Monogr. Röm.-Germ. Zentralmus. Mainz 40, 2, I-VIII + 419-698 + pls. 68-131 + 1 suppl., Habelt, Bonn.

(ed.) 2001b: Das Pleistozän von Untermaßfeld bei Meiningen (Thüringen), Part 3. Monogr. Röm.-Germ. Zentralmus. Mainz 40, 2, I-VI + 699-1030 + pls. 132-151 + 15 suppl., Habelt, Bonn.

2001c: Die Fortführung der Forschungsarbeiten zur unterpleistozänen Komplexfundstelle Untermaßfeld in den Jahren 1989 - 1996. In: Kahlke, R.-D. (ed.): Das Pleistozän von Untermaßfeld bei Meiningen (Thüringen), Part 2, Monogr. Röm.-Germ. Zentralmus. Mainz 40, 2, 419-432, Habelt, Bonn.

2001d: Verbesserte geologische Standardprofile zur unterpleistozänen Komplexfundstelle Untermaßfeld. In: Kahlke, R.-D. (ed.): Das Pleistozän von Untermaßfeld bei Meiningen (Thüringen), Part 2, Monogr. Röm.-Germ. Zentralmus. Mainz 40, 2, 433-445 + 1 suppl., Habelt, Bonn.

2001e: Schädelreste von *Hippopotamus* aus dem Unterpleistozän von Untermaßfeld. In: Kahlke, R.-D. (ed.): Das Pleistozän von Untermaßfeld bei Meiningen (Thüringen), Part 2, Monogr. Röm.-Germ. Zentralmus. Mainz 40, 2, 483-500 + pls. 77-78, Habelt, Bonn.

2001f: Ein Hinweis auf Herbivoren-Osteophagie im Unterpleistozän von Untermaßfeld sowie Bemerkungen zur Gabelbildung an Knochenenden. In: Kahlke, R.-D. (ed.): Das Pleistozän von Untermaßfeld bei Meiningen (Thüringen), Part 3, Monogr. Röm.-Germ. Zentralmus. Mainz 40, 3, 897-903 + pl. 147, Habelt, Bonn.

2001g: Die unterpleistozäne Komplexfundstelle Untermaßfeld – Zusammenfassung des Kenntnisstandes sowie synthetische Betrachtungen zu Genesemodell, Paläoökologie und Stratigraphie. In: Kahlke, R.-D. (ed.): Das Pleistozän von Untermaßfeld bei Meiningen (Thüringen), Part 3, Monogr. Röm.-Germ. Zentralmus. Mainz 40, 3, 931-1030 + suppl. 1-15, Habelt, Bonn.

2002a: The late Lower Pleistocene (Epivillafranchian) faunal site of Untermassfeld. In: Meyrick, R. A. & Schreve, D. C. (eds.): The Quaternary of Central Germany (Thuringia & Surroundings), Field Guide, 97-114, Quaternary Research Association, London.

2002b: Bericht zu den bodendenkmalpflegerischen Maßnahmen an der unterpleistozänen Komplexfundstelle Untermaßfeld im Jahr 2002. Unpublished excavation report, 1-15 + suppl., Weimar.

2003a: Reiche Neufunde in Untermaßfeld: Die Grabungskampagne 2002. Natur u. Museum 133, 2, 54-57, Frankfurt a. M.

2003b: Bericht zu den bodendenkmalpflegerischen Maßnahmen an der unterpleistozänen Komplexfundstelle Untermaßfeld im Jahr 2003. Unpublished excavation report, 1-19, Weimar.

2004a: Quartäre Großsäugetiere. Natur u. Museum 134, 11, 368-369, Frankfurt a. M.

2004b: Bericht zu den bodendenkmalpflegerischen Maßnahmen an der unterpleistozänen Komplexfundstelle Untermaßfeld im Jahr 2004. Unpublished excavation report, 1-20 + suppl., Weimar.

Kahlke, R.-D. & Gaudzinski, S. 2005: The blessing of a great flood: differentiation of mortality patterns in the large mammal record of the Lower Pleistocene fluvial site of Untermassfeld (Germany) and its relevance for the interpretation of faunal assemblages from archaeological sites. J. Archaeol. Sci. 32, 1202-1222, London etc.

Kahlke, R.-D. & Kierdorf, U. 1997: Diffuse idiopathische skelettale Hyperostose bei einem Bären (*Ursus* sp.) aus dem Unterpleistozän von Untermaßfeld bei Meiningen (Südthüringen, Deutschland). Quartär 47/48, 159-175, Bonn.

Kästner, H., Seidel, G. & Wiefel, H. 1995: Regionalgeologische Stellung und Gliederung. In: Seidel, G. (ed.): Geologie von Thüringen, 13-21, Schweizerbart, Stuttgart.

Keiler, J.-A. 1995: Bergung und Präparation pleistozäner Wirbeltierreste unter Berücksichtigung des Fossilmaterials der Komplexfundstelle Untermaßfeld / Südthüringen. Restaurierung und Museumstechnik 12, 1-31 + pls. 1-24, Weimar.

2001a: Die Polyvinyl-Tränkungsmethode für fossile Knochen. Der Präparator 12, 2, 49-56, Nierstein.

2001b: Die Präparation und Konservierung des

Fossilmaterials aus dem Unterpleistozän von Untermaßfeld. In: Kahlke, R.-D. (ed.): Das Pleistozän von Untermaßfeld bei Meiningen (Thüringen), Part 2, Monogr. Röm.-Germ. Zentralmus. Mainz 40, 2, 447-459 + pls. 68-71, Habelt, Bonn.

2001c: Die Koprolithen aus dem Unterpleistozän von Untermaßfeld. In: Kahlke, R.-D. (ed.): Das Pleistozän von Untermaßfeld bei Meiningen (Thüringen), Part 2, Monogr. Röm.-Germ. Zentralmus. Mainz 40, 2, 691-698 + pls. 128-131, Habelt, Bonn.

Kierdorf, U. & Kahlke, R.-D. 2004: Pathological findings on remains of *Hippopotamus amphibius antiquus* Desmarest, 1822 (Hippopotamidae, Mammalia) from the late Lower Pleistocene site of Untermaßfeld. In: Maul, L. C. & Kahlke, R.-D. (eds.): Late Neogene and Quaternary biodiversity and evolution: Regional developments and interregional correlations. Conference Volume, 18th International Senckenberg Conference, VI International Palaeontological Colloquium in Weimar, Weimar (Germany), 25th - 30th April, 2004, Terra Nostra, Schriften der Alfred-Wegener-Stiftung 2004/2, 137-138, Berlin.

Kiladze, G. 2003: Lynx (*Lynx issiodorensis*) from Dmanisi fauna. In: Kopaliani, J. (ed.): Dmanisi IV, 131-138, Metsniereba, Tbilisi (in Georgian, Russian and English summary).

Kleinschmidt, A. 1965: Die Mechanik der Sedimentbewegung in kleinen Flußrinnen. Ein Beitrag zur analytischen Sediment-Gefügekunde (Stratonomie). Jh. Ver. vaterl. Naturkde. Württemberg 120, 126-184, Stuttgart.

Klingel, H. 1995: Zoff in Hippo-Hills. BBC Wildlife Magazin 1995, 8, 6-15, London.

Kočenov, V. G. & Kožamkulova, B. S. 1988: Novaâ pozdnepliocenovaâ lošad' iz Aktogaâ (Ilijskaâ vpadina, r. Čaryn). In: Mezregional'noe sopostavlenie mezozoj-kajnozojskih faun i flor Kazahstana, Materialy ist. fauny flory Kazahstana 10, 99-120, Nauka, Alma-Ata (in Russian).

1990: Slon Gromova ûgo-vostočnoj časti Kazahstana i ego stratigrafičeskaâ priuročennost'. In: Fauna pozvonočnyh i flora mezozoâ i kajnozoâ Kazahstana 11, 75-80, Nauka, Alma-Ata (in Russian).

Koenigswald, W. v. 1991: Exoten in der Großsäuger-Fauna des letzten Interglazials von Mitteleuropa. Eiszeitalter u. Gegenwart 41, 70-84, Hannover.

1992: Zur Ökologie und Biostratigraphie der beiden pleistozänen Faunen von Mauer bei Heidelberg. In: Beinhauer, K. W. & Wagner, G. A. (ed.): Schichten von Mauer – 85 Jahre *Homo erectus heidelbergensis*, 101-110, Reiß-Museum, Edition Braus, Mannheim.

1997: Die fossilen Säugetiere aus den Sanden von Mauer. In: Wagner, G. A. & Beinhauer, K. W. (eds.): *Homo heidelbergensis* von Mauer – Das Auftreten des Menschen in Europa, 215-240, Heidelberger Verlagsanst., Heidelberg.

Koenigswald, W. v. & Heinrich, W.-D. 1999: Mittelpleistozäne Säugetierfaunen aus Mitteleuropa – der Versuch einer biostratigraphischen Zuordnung. In: Schrenk, F. & Gruber, G. (eds.): Current Research 2 (Plio-Pleistocene Mammalian Evolution), Kaupia, Darmstädter Beitr. Naturgesch. 9, 53-112, Darmstadt.

Koenigswald, W. v. & Kolfschoten, T. v. 1996: The *Mimomys-Arvicola* boundary and the enamel thickness quotient (SDQ) of *Arvicola* as stratigraphic markers in the Middle Pleistocene. In: Turner, C. (ed.): The early Middle Pleistocene in Europe, Proceedings of the SEQS Cromer Symposium Norwich / United Kingdom, 3-7 September 1990, 211-226, Balkema, Rotterdam, Brookfield.

Koenigswald, W. v. & Tobien, H. 1987: Bemerkungen zur Altersstellung der pleistozänen Mosbach-Sande bei Wiesbaden. Geol. Jb. Hessen 115, 227-237, Wiesbaden.

Kolfschoten, T. v. 1990a: The Early Biharian Mammal Fauna from Bavel and Dorst-Surae. Quartärpaläontologie 8, 265-272, Berlin.

1990b: The evolution of the mammal fauna in the Netherlands and the Middle Rhine area (Western Germany) during the late Middle Pleistocene. Meded. Rijks Geol. Dienst 43, 3, 1-69, Haarlem.

1994: Mammalian remains in a palaeolithic context. In: Milliken, S. & Peretto, C. (eds.): Archaeology, Methodology and the Organization of the Research, Acts of the Round Table, Isernia 27 May 1994, 19-35, A.B.A.C.O., Forli.

2001: Pleistocene Mammals from the Netherlands. Boll. Soc. Paleont. Ital. 40, 2, 209-215, Modena.

Kolfschoten, T. v. & Essen, H. v. 2004: Palaeozoological hertiage from the bottom of the North Sea. In: Flemming, N. C. (ed.): Submarine prehistoric archaeology of the North Sea. Research priorities and collaboration with industry, CBA Res. Rep. 141, 70-80, York.

Kolfschoten, T. v. & Laban, C. 1995: Pleistocene terrestrial mammal faunas from the North Sea. Meded. Rijks Geol. Dienst 52, 135-151, Haarlem.

Kolfschoten, T. v. & Turner, E. 1996: Early Middle Pleistocene mammalian faunas from Kärlich and Miesenheim I and their biostratigraphical implications. In: Turner, C. (ed.): The early Middle Pleistocene in Europe, Proceedings of the SEQS Cromer Symposium Norwich / United Kingdom, 3-7 September 1990, 227-253, Balkema, Rotterdam, Brookfield.

Kolfschoten, T. v. & Vervoort-Kerkhoff, Y. 1999a: The Pleistocene and Holocene mammalian assemblages from the Maasvlakte near Rotterdam (The Netherlands), with special reference to the Proboscidea. In: Reumer, J. W. F. & Vos, J. de (eds.): Official Conference Papers, Volume of Abstracts, Excursion Guide, 2nd International Mammoth Conference, 200 Years of Mammoth Research, May 16 - 20, 1999, Natuurmuseum Rotterdam, The Netherlands, 89-99, Natuurmuseum Rotterdam, Rotterdam.

1999b: The Pleistocene and Holocene Mammalian assemblages from the Maasvlakte near Rotterdam (the Netherlands), with special reference to the Ovibovini *Soergelia minor* and *Praeovibos* cf. *priscus*. In: Reumer, J. W. F. & Vos, J. de (eds.): Elephants have a snorkel! Papers in honour of Paul Y. Sondaar, Deinsea 7, 369-381, Rotterdam.

Kostopoulos, D. S. 1997: The Plio-Pleistocene artiodactyls (Vertebrata, Mammalia) of Macedonia. 1. The fossiliferous site "Apollonia-1", Mygdonia basin of Greece. Geodiversitas 19, 4, 845-874, Paris.

1998: *Parastrepsiceros koufosi* n. sp. (Mammalia: Bovidae); note on the possible presence of a *Prostrepsiceros* descendant in the latest Pliocene of northern Greece. Acta zool. cracov. 41, 1, 101-109, Kraków.

Kostopoulos, D. S. & Athanassiou, A. S. 1997: Les Gazelles du Pliocène moyen-terminal de la Grèce continentale (Macédoine, Thessalie). N. Jb. Geol. Paläont. Abh. 205, 3, 413-430, Stuttgart.

Kostopoulos, D. S. & Koufos, G. D. 1994: The Plio-Pleistocene artiodactyls of Macedonia (Northern Greece) and their biostratigraphic significance; preliminary report. C. R. Acad. Sci. Paris 318, Sér. II, 1267-1272, Paris.

Koufos, G. D. 1986: The presence of *Sus strozzii* in the Villafranchian (Villanyian) of Macedonia (Greece). Paläont. Z. 60, 3/4, 341-351, Stuttgart.

1992: The Pleistocene carnivores of the Mygdonia basin (Macedonia, Greece). Ann. Paléont. 78, 4, 205-257, Paris.

1993: Late Pliocene carnivores from western Macedonia (Greece). Paläont. Z. 67, 3/4, 357-376, Stuttgart.

Koufos, G. D. & Kostopoulos, D. S. 1997: Biochronology and succession of the Plio-Pleistocene macromammalian localities of Greece. In: Aguilar, J.-P., Legendre, S. & Michaux, J. (eds.): Actes du Congès BiochroM'97, Mém. Trav. E. P. H. E., Inst. Montpellier 21, 619-634, Montpellier.

Koufos, G. D., Kostopoulos, D. S. & Sylvestrou, I. A. 1997: *Equus apolloniensis* n. sp. (Mammalia, Equidae) from the latest Villafranchian locality of Apollonia, Macedonia, Greece. Paleont. i Evol. 30-31, 49-76, Sabadell.

Koufos, G. D. & Melentis, J. K. 1983: New data from the Villafranchian mammal locality of Gerakarou (Macedonia - Greece). Proc. Acad. Athens 58, 185-191, Athens.

Koufos, G., Syrides, G. & Koliadimou, K. 1989: A new Pleistocene mammal locality from Macedonia (Greece). Contribution to the study of Villafrancian (Villangian) [*sic!*] in Central Macedonia. Bull. Geol. Soc. Greece 23, 2, 113-124, Athens.

Kowalski, K. (ed.) 1989: Historia i ewolucja ladowej fauny Polski. History and Evolution of the terrestrial fauna of Poland. Folia Quaternaria 59-60, 1-278, Kraków.

Kožamkulova, B. S. 1969: Antropogenovaâ iskopaemaâ teriofauna Kazahstana. 149 p.+ 34 pls., Nauka, Alma-Ata (in Russian).

1977: Zoogeografičeskij analis sostava i raspredeleniâ antropogenovoj fauny Kazahstana v prostranstve i vremeni. In: Kožamkulova, B. S., Tleuberdina, P. A., Savinov. P. F. & Orlovskaâ, E. R.: Mezokajnozojskaâ fauna i flora severo-zapadnogo Kazahstana, Materialy ist. fauny flory Kazahstana 7, 89-104, Nauka, Alma-Ata (in Russian).

1990: Teriokompleksy pozdnego pliocena i rannego plejstocena Kazahstana i ih analogi v vostočnoj Evrope i asiatskoj časti SSSR. In: Fauna pozvonočnyh i flora mezozoâ i kajnozoâ Kazahstana 11, 5-17, Nauka, Alma-Ata (in Russian).

Kožamkulova, B. S., Savinov, P. F., Tût'kova, L. A. & Pak, T. K. 1987: Pliocennovye mlekopitaûšie uročiša Aktogaj. In: Mezozojskaâ i kajnozojskaâ fauna i flora zentral'nogo Kazahstana, ûznogo Altaâ i severnogo Tân'-Šanâ, Materialy ist. fauny flory Kazahstana 9, 82-120, Nauka, Alma-Ata (in Russian).

Kretzoi, M. 1937/38: Die Raubtiere von Gombaszög nebst einer Übersicht der Gesamtfauna. (Ein Beitrag zur Stratigraphie des Altquartaers). Ann. Mus. Nat. Hung. 31, Pars Min. Geol. Palaeont., 88-157 + pls. 1-3, Budapest.

1941a: Weitere Beiträge zur Kenntnis der Fauna von Gombaszög. Ann. Mus. Nat. Hung. 34, Pars Min. Geol. Palaeont., 105-139 + pl. 5, Budapest.

1941b: Die unterpleistozäne Säugetierfauna von Betfia bei Nagyvárad. Földt. Közl. 71, 7-12, 308-335, Budapest.

1954: Jelentés a kislángi kalabriai (villafrankai) fauna feltárásáról. (Bericht über die Calabrische (Villafranchische) Fauna von Kisláng). Kom. Fejér. Földt. Int. Évi. Jel. 1, 213-264, Budapest.

1956: Die altpleistozänen Wirbeltier-Faunen des Villányer Gebirges. Geol. Hung., Ser. Palaeont. 27, 1-264, Budapest.

1965: Die Nager und Lagomorphen von Voigtstedt in Thüringen und ihre chronologische Aussage. Paläont. Abh. A, Paläozool, II, 2/3, 585-660 + pls. XXXVIII-XXXIX, Berlin.

1990: Vertebrate fauna of the archaeological site. In: Kretzoi, M. & Dobosi, V. T. (eds.): Vértesszölös. Man, Site and Culture, 231-247, Akad. Kiadó, Budapest.

Krolopp, E. 1997: Die Molluskenreste aus dem Unterpleistozän von Untermaßfeld. In: Kahlke, R.-D. (ed.): Das Pleistozän von Untermaßfeld bei Meiningen (Thüringen), Part 1, Monogr. Röm.-Germ. Zentralmus. Mainz 40, 1, 71-75 + pl. 19, Habelt, Bonn.

Kruuk, H. 1972: The Spotted Hyena. A Study of Predation and Social Behavior. xvi + 335 p., Univ. Chicago Press, Chicago, London.

Kurtén, B. 1968: Pleistocene Mammals of Europe. 317 p., Weidenfeld & Nicolson, London.

Kuss, S. E. 1955: Die stratigraphische Verteilung der altpleistocänen Fauna aus dem Tonlager von Jockgrim in der Pfalz. Beitr. naturkdl. Forsch. Südwestdeutschland 14, 2, 93-100, Freiburg i. Br.

1957: Altpleistozäne Reste des *Hippopotamus antiquus* DESMAREST vom Oberrhein. Jh. geol. Landesamt Baden-Württemberg 2, 299-331, Freiburg i. Br.

Kuz'mina, I. E. & Sablin, M. V. 1991: O novoj nahodke iskopaemyh ostatkov *Equus stenonis* Cocchi v Zakavkaz'e. In: Baryšnikov, G. F. & Kuz'mina, I. E. (eds.): Paleoteriologičeskie issledovaniâ fauny SSSR, Akad. Nauk SSSR, Trudy Zool. Inst. 238, 61-67, Sankt-Peterburg (in Russian, English summary).

Latham, A. G. & Schwarcz, H. P. 1990: Magnetic polarity of travertine samples from Vértesszölös. In: Kretzoi, M. & Dobosi, V. T. (eds.): Vértesszölös. Man, Site and Culture, 553-555, Akad. Kiadó, Budapest.

Laws, R. M. 1968: Dentition and ageing of the Hippopotamus. E.-Afr. Wildl. J. 6, 19-52, Oxford.

Lebedeva, N. A. 1978: Korrelâziâ antropogenovyh tolš Ponto-Kaspiâ. 135 p., Nauka, Moskva (in Russian).

Lindsay, E. H. 1990: The setting. In: Lindsay, E. H., Fahlbusch, V. & Mein, P. (eds.): European Neogene Mammal Chronology, NATO ASI Ser. A, Life Sci. 180, 1-14, Plenum Press, New York, London.

Lister, A. M. 1993: The stratigraphical significance of deer species in the Cromer Forest-bed Formation. J. Quatern. Sci. 8, 2, 95-108, New York etc.

1996a: Evolution and taxonomy of Eurasian mammoths. In: Shoshani, J. & Tassy, P. (eds.): The Proboscidea, Evolution and Palaeoecology of Elephants and Their Relatives, 203-213, Oxford Univ. Press, Oxford etc.

1996b: The stratigraphical interpretation of large mammal remains from the Cromer Forest-bed Formation. In: Turner, C. (ed.): The early Middle Pleistocene in Europe, Proceedings of the SEQS Cromer Symposium, Norwich / United Kingdom, 3-7 September 1990, 25-44, Balkema, Rotterdam, Brookfield.

1998: The age of Early Pleistocene mammal faunas from the 'Weybourne Crag' and Cromer Forest-bed Formation (Norfolk, England). In: Kolfschoten, T. v. & Gibbard, P. L. (eds.): The Dawn of the Quaternary, Proceedings of the SEQS-EuroMam symposium 1996, Meded. Nederl. Inst. Toegepaste Geowetenschappen TNO 60, 271-280, Haarlem.

Lister, A. M., McGlade, J. M. & Stuart, A. J. 1990: The Early Middle Pleistocene vertebrate fauna from Little Oakley, Essex. Phil. Trans. R. Soc. London, B, 328, 359-385, London.

Lister, A. M., Sher, A. V., Essen, H. v. & Wei, G. 2005: The pattern and process of mammoth evolution in Eurasia. Quaternary International 126-128, 49-64, Amsterdam etc.

Lumley, H. de, Kahlke, H.-D., Moigne, A.-M. & Moullé, P.-E 1988: Les faunes de grands mammifères de la grotte du Vallonnet Roquebrune-Cap-Martin, Alpes-Maritimes. L`Anthropologie 92, 2, 465-496, Paris.

Lüps, P. & Wandeler, A. I. 1993: *Meles meles* (Linnaues, 1758) – Dachs. In: Stubbe, M. & Krapp, F. (eds.): Handbuch der Säugetiere Europas, Vol. 5: Raubsäuger – Carnivora (Fissipedia), Part II: Mustelidae 2, Viverridae, Herpestidae, Felidae, 856-906, Aula-Verl., Wiesbaden.

Made, J. v. d. 1998: Ungulates from Gran Dolina (Atapuerca, Burgos, Spain). Quaternaire 9, 4, 267-281, Paris.

1999: Ungulates from Atapuerca TD6. J. Human Evol. 37, 3/4, 389-413, London etc.

2001: Les Ongulés d'Atapuerca. Stratigraphie et biogéographie. L'Anthropologie 105, 95-113, Paris.

Mais, K. & Rabeder, G. 1984: Das große Höhlensystem im Pfaffenberg bei Bad Deutsch-Altenburg (Niederösterreich) und seine fossilen Faunen. Die Höhle 35, 3/4, 213-230, Wien.

Majsuradze, G., Pavlenisvili, E., Schmincke, H.-U. & Sologasvili, D. 1989: Paläomagnetik und Datierung der Basaltlava. In: Dzaparidze, V., Bosinski, G., Bugianisvili, T., Gabunia, L., Justus, A., Klopotovskaja, N., Kvavadse E., Lordkipanidze, D., Majsuradze, G., Mgeladze, N., Nioradze, M., Pavenisvili, E., Schmincke, H.-U., Sologasvili, D., Tusabramisvili, D., Tvalcrelidze, M. & Vekua, A.: Der altpaläolithische Fundplatz Dmanisi in Georgien (Kaukasus), Jb. Röm.-Germ. Zentralmus. Mainz 36, 74-76, Habelt, Bonn.

Malez, M. 1959: A new Old-Pleistocene fauna in Dalmatia. Bulletin scientifique, Conseil des Academies des Sciences et des Arts de la RSF de Yugoslavie 5, 1, 13-14, Zagreb.

1960: Etruskisches Nashorn – *Dicerorhinus etruscus* (Falconer) – aus altpleistozänen Brekzien der Halbinsel Marjan bei Split (Dalmatien). Anthropos, supl. 1960, Mammalia pleistocaenica I, 115-125, Brno.

1961: Staropleistocenska fauna kostane brece poluotoka Marjana kod Splita. Die altpleistozäne Brekzienfauna der Halbinsel Marjan bei Split. Palaeontologia Jugoslavica 4, 1-43, Zagreb.

1965: Der altpleistozäne Fundort Dubci in Mitteldalmatien. Bulletin scientifique, Conseil des Academies des Sciences et des Arts de la RSF de Yugoslavie 10, 12, 418-420, Zagreb.

1967: Donjopleistocenska fauna kostane brece kod sela Dubci u Dalmaciji. Die altpleistozäne Knochenbrekzienfauna beim Dorfe Dubci in Dalmatien. Jugoslav. Akad. Znanosti Umjetnosti, otd. prir. nauke, 345, 53-101, Zagreb.

1968: Die altpleistozänische Vertebratenfauna in dem Gebiete des Dinarischen Karstes. Bulletin scientifique, Conseil des Academies des Sciences et des Arts de la RSF de Yugoslavie, A, 13, 5-6, 152, Zagreb.

1969: Die altpleistozänen Wirbeltierfaunen aus den Höhlen von Kroatien. In: V. Int. Kongr. Speläologie Stuttgart 1969, Abh. 4, Biospeläologie, B 21/1-6, München.

1975: O znacenju otkrica ostatka roda *Homo* u naslagama Vilafranka Šandalje I kod Pule. On the significance of the genus *Homo* discovery in the Villafranchian sediments of Sandalja I near

Pula. RAD Jugoslav. akad. znanosti umjetnosti 371, 181-201, Zagreb.

1986: Die quartären Vertebraten-Faunen in der SFR Jugoslawien. Quartärpaläontologie 6, 101-117, Berlin.

Malez, M., Forstén, A. & Lenardic, J. 1992: Fossil horses (Mammalia, Equidae) from the bone breccias of Croatia, northern Balcan. Paläont. Z. 66, 3/4, 369-385, Stuttgart.

Martinez, B., Espigares, M. P. & Ros, S. 2003: Estudio preliminar de las asociaciones de grandes mamíferos de Fuente Nueva-3 y Barranco León-5 (Orce, Granada, Espana) (Informe de las campanas de 1999-2002). In: Toro, I., Agusti, J. & Martinez-Navarro, B. (eds.): El Pleistoceno Inferior de Barranco León y Fuente Nueva 3, Orce (Granada), Memoria científica campanas 1999-2002, 115-136, Consejeria de Cultura, Sevilla.

Martinez Navarro, B. & Palmqvist, P. 1995: Presence of the African Machairodont *Megantereon whitei* (Broom, 1937) (Felidae, Carnivora, Mammalia) in the Lower Pleistocene Site of Venta Micena (Orce, Granada, Spain), with some considerations on the Origin, Evolution and Dispersal of the Genus. J. Archaeol. Sci. 22, 569-582, London etc..

1996: Presence of the African Saber-toothed Felid *Megantereon whitei* (Broom, 1937) (Mammalia, Carnivora, Machairodontinae) in Apollonia-1 (Mygdonia Basin, Macedonia, Greece). J. Archaeol. Sci. 23, 869-872, London etc.

1999: Venta Micena (Orce, Granada, Spain): Human activity in a hyena den during the Lower Pleistocene. In: The role of early humans in the accumulation of European Lower and Middle Palaeolithic bone assemblages, Monogr. Röm.-Germ. Zentralmus. Mainz 42, 57-71, Habelt, Bonn.

Martinez Navarro, B., Palmqvist, P. & Arribas, A. 1998: The paleontological site at Venta Micena (Orce, Guadix-Baza Basin, Province of Granada, Spain). In: Agusti, J., Oms, O. & Martin-Suárez, E. (eds.): The Plio-Pleistocene Vertebrate Succession of the Guadix-Baza Basin (SE Spain), Field Guide, 23-24, European Quaternary Mammal Research Association (INQUA-SEQS), Granada.

Martinez Navarro, B., Turq, A., Agusti Ballester, J. & Oms, O. 1997: Fuente Nueva-3 (Orce, Granada, Spain) and the first human occupation of Europe. J. Human Evol. 33, 611-620, London etc.

Masini, F., Abbazzi, L., Ferretti, M. & Mazza, P. 1995: Fauna a mammiferi terrestri dalle "Sabbie Gialle" quaternarie di Imola. In: Pacciarelli, M. & Var, G. B. (eds.): Catalogo delle collezioni, La Collezione Scarabelli, I. Geologia, 314-318, Musei Civici di Imola, Imola.

Masini, F., Ficcarelli, G. & Torre, D. 1994: Late Villafranchian and the Earliest Galerian mammal faunas from some intermontane basins of north-central Italy. Mem. Soc. Geol. Ital. 48, 381-389, Pisa.

Masini, F., Sala, B. & Vorlicek, P. A. 1995: Late Villafranchian mammals from a karst fissure at Alonte (Berici Hills, Vicenza, Northern Italy). Il Quaternario 8, 2, 443-448, Verona.

Maul, L. 1997: Nachweis eines Zungenbeinfragmentes einer Schildkröte (Emydini, Emydidae, Testudinata) aus dem Unterpleistozän von Untermaßfeld. In: Kahlke, R.-D. (ed.): Das Pleistozän von Untermaßfeld bei Meiningen (Thüringen), Part 1, Monogr. Röm.-Germ. Zentralmus. Mainz 40, 1, 89-94, Habelt, Bonn.

2001a: Die Kleinsäugerreste (Insectivora, Lagomorpha, Rodentia) aus dem Unterpleistozän von Untermaßfeld. In: Kahlke, R.-D. (ed.): Das Pleistozän von Untermaßfeld bei Meiningen (Thüringen), Part 3, Monogr. Röm.-Germ. Zentralmus. Mainz 40, 3, 783-887 + pls. 144-145, Habelt, Bonn.

2001b: Nagespuren von Kleinsäugern an Knochenmaterial aus der unterpleistozänen Komplexfundstelle Untermaßfeld. In: Kahlke, R.-D. (ed.): Das Pleistozän von Untermaßfeld bei Meiningen (Thüringen), Part 3, Monogr. Röm.-Germ. Zentralmus. Mainz 40, 3, 905-916 + pls. 148-151, Habelt, Bonn.

Maul, L., Masini, F., Abbazzi, L. & Turner, A. 1998: The use of different morphometric data for absolute age calibration of some South- and Middle European arvicolid populations. Palaeontograph. Ital. 85, 111-151, Pisa.

Maul, L. C., Rekovets, L., Heinrich, W.-D., Keller, T. & Storch, G. 2000: *Arvicola mosbachensis* (SCHMIDTGEN 1911) of Mosbach 2: a basic sample for the early evolution of the genus and a reference for further biostratigraphical studies.

In: Storch G. & Weddige K. (eds.): Advances in Vertebrate Palaeontology, Senckenbergiana lethaea 80, 1, 129-147, Frankfurt a. M.

Mäuser, M. 1987: Geologische und paläontologische Untersuchungen an der altpleistozänen Säugetier-Fundstelle Würzburg-Schalksberg. Münchner Geowiss. Abh., A, 11, 1-78, München.

Mayhew, D. F. 1978: Reinterpretation of the extinct beaver *Trogontherium* (Mammalia, Rodentia). Phil. Trans. R. Soc. London, 281, B 983, 407-438, London.

Mazo, A. V., Sesé, C., Ruiz Bustos, A. & Peña, J. A. 1985: Geologia y paleontologia de los yacimientos plio-pleistocenos de Huescar (Depresion de Guadix-Baza, Granada). Estud. geol. 41, 467-493, Madrid.

Mazza, P. 1991: Interrelations between Pleistocene hippopotami of Europe and Africa. Boll. Soc. Paleont. Ital. 30, 2, 153-186, Modena.

1995: New evidence on the Pleistocene hippopotamuses of Western Europe. Geol. Rom. 31, 61-241, Roma.

Mazza, P. & Rustioni, M. 1992: Morphometric revision of the Eurasian species *Ursus etruscus* Cuvier. Palaeontograph. Ital. 79, 101-146, Pisa.

Mazza, P., Sala, B. & Fortelius, M. 1993: A small latest Villafranchian (late Early Pleistocene) rhinoceros from Pietrafitta (Perugia, Umbria, Central Italy), with notes on the Pirro and Westerhoven rhinoceroses. Palaeontograph. Ital. 80, 25-50, Pisa.

Meijer, T. 1996: References of relevant publications about Pliocene and Lower-Pleistocene deposits in The Netherlands. 18 p., INQUA-SEQS '96, The Dawn of the Quaternary, 16 - 21 June 1996, Kerkrade – The Netherlands, Geol. Survey of The Netherlands, Kerkrade.

Meulen, A. J. v. d. & Kolfschoten, T. v. 1986: Review of the Late Turolian to Early Biharian mammal faunas from Greece and Turkey. Mem. Soc. Geol. It. 31, 201-211, Firenze.

Miller, G. J. 1969: A study of cuts, grooves, and other marks on recent and fossil bone. I. Animal tooth marks. Tebiwa 12, 20-26, Pocatello.

Mohr, E. 1965: Altweltliche Stachelschweine. 164 p., Ziemsen, Wittenberg.

Moigne, A.-M. 1984: The giant muskox of the Arago Cave, Tautavel, France. Biol. Pap. Univ. Alaska Spec. Rep. 4, 145-147, Fairbanks.

Moigne, A.-M., Belda, V., Briki, D., Kacimi, S., Lacombat, F., Moutoussamy, J., Quiles, J., Rivals, F. & Testu, A. 2000: Les faunes de grands mammifères du Pléistocène moyen de la Caune de l'Arago (complex stratigraphique moyen), à Tautavel. Leur place dans le cadre biostratigraphique et paléoclimatique du Quaternaire. Les horizons biostratigraphiques Arago CM I, CM II et CM III. In: Les premiers habitants de l'Europe, Résumés des Communications, 57-58, Tautavel.

Mol, D. 1998: Uit de literatuur. Kahlke, R. D. (ed), 1997. Das Pleistozän von Untermaßfeld bei Meiningen (Thüringen). Teil I. - Römisch-Germanisches Zentralmuseum Mainz, Monographien, Band 40, viii + 418 pp., 67 platen, 166 tekeningen en foto's in tekst. Verlag Römisch-Germanisches Zentralmuseum Mainz. Cranium 15, 1, 41-43, Utrecht.

Mol, D., Bergh, G. D. v. d. & Vos, J. de 1999: Fossil proboscideans from The Netherlands, the North Sea and the Oosterschelde Estuary. In: Haynes, G., Klimowicz, J. & Reumer, J. W. F. (eds.): Mammoths and the Mammoth Fauna: Studies of an Extinct Ecosystem, Proceedings of the First International Mammoth Conference, St. Petersburg, Russia, October 16 - 21, 1996, Deinsea 6, 119-145, Rotterdam.

Moullé, P.-É. 1990: Les cervidés de la grotte du Vallonnet (Roquebrune-Cap-Martin, Alpes-Maritimes). Quaternaire 1, 3-4, 193-196, Paris.

1997-1998: Les grands mammifères de la grotte du Vallonnet (Roquebrune-Cap-Martin, Alpes-Maritimes). Synthèse des études antérieures et nouvelles déterminations. Bull. Mus. Anthrop. préhist. Monaco 39, 29-36, Monaco.

1999: Paléontologie des grands mammifères de la grotte du Vallonnet (Roquebrune Cap-Martin, Alpes-Maritimes, France). In: Gibert, J., Sánchez, F., Gibert, L. & Ribot, F. (eds.): Los Hominidos y su entorno en el Pleistoceno inferior y medio de Eurasia, Actas del Congreso Internacional de Paleontologia Humana, Orce 1995, 571-582, Museo Prehistoria y Paleontologia "J. Gibert", Orce.

Moullé, P.-E., Echassoux, A., Moigne, A.-M., Palombo, M.-R., Caloi, L., Kahlke, R.-D., Vekua, A. & Lordkipandize, D. 2000: Les faunes de la fin du Pléistocène inférieur de la grotte du Vallonnet (Alpes-Maritimes, France), de Redicicoli (Latium, Italie), de Untermaßfeld (Allemagne) et de Akhalkalaki (Géorgie): L'horizon biostratigraphique du Vallonnet. In: Les premiers habitants de l'Europe, Résumés des Communications, 50-51, Tautavel.

Moyà-Solà, S. 1987: Los bóvidos (Artiodactyla, Mammalia) del yacimiento del Pleistoceno inferior de Venta Micena (Orce, Granada, España). Paleont. i Evol., Mem. Esp. 1, 181-236, Sabadell.

Moyà-Solà, S., Agusti, J., Gibert, J. & Vera, J. A. (eds.) 1987: Geologia y Paleontologia del Pleistoceno inferior de Venta Micena. Paleont. i Evol., Mem. Esp. 1, 1-295, Sabadell.

Musil, R. (ed.) 1971: Stránská skála I. 1910 - 1945. Studia Musei Moraviae, Anthropos 20 (N. S. 12), 1-204, Brno.

(ed.) 1995a: Stránská skála Hill. Excavation of open-air sediments 1964 - 1972. Anthropos 26 (N. S. 18), 1-213, Brno.

1995b: Large fauna of talus cones at the Stránská Skála Hill. In: Musil, R. (ed.) 1995: Stránská Skála Hill, Excavation of open-air sediments 1964 - 1972, Anthropos 26 (N. S. 18), 65-83, Brno.

2001a: Die Equiden-Reste aus dem Unterpleistozän von Untermaßfeld. In: Kahlke, R.-D. (ed.): Das Pleistozän von Untermaßfeld bei Meiningen (Thüringen), Part 2, Monogr. Röm.-Germ. Zentralmus. Mainz 40, 2, 557-587 + pls. 92-94, Habelt, Bonn.

2001b: Die Ursiden-Reste aus dem Unterpleistozän von Untermaßfeld. In: Kahlke, R.-D. (ed.): Das Pleistozän von Untermaßfeld bei Meiningen (Thüringen), Part 2, Monogr. Röm.-Germ. Zentralmus. Mainz 40, 2, 633-658 + pls. 114-120, Habelt, Bonn.

Mussi, M. 1995: The earliest occupation of Europe: Italy. In: Roebroeks, W. & Kolfschoten, T. v. (eds.): The Earliest Occupation of Europe, Proceedings of the European Science Foundation Workshop at Tautavel (France), 1993, 27-49, Univ. Leiden, Leiden.

Nikiforova, K. V. (ed.) 1982: Guidebook for excursions A-7, C-7. Moldavia and the Odessa Region. 45 p., INQUA, XI Congress, Moscow.

Nikiforova, K. V., Belâeva, E. I., Vangengejm, E. A., Konstantinova, N. A. & Negadaev-Nikonov, K. N. (eds.) 1971: Plejstocen Tiraspolâ. 187 p., Stiinza, Kišinev (in Russian).

Nikiforova, K. V. & Vangengejm, E. A. (eds.) 1988: Biostratigrafiâ pozdnego pliocena - rannego plejstocena Tadzikistana (po faune mlekopitaûših). 126 p., Nauka, Moskva (in Russian).

Nobis, G. 1981: *Equus mosbachensis* v. Reichenau aus Ablagerungen des cromerzeitlichen Mains von Randersacker bei Würzburg. Quartärpaläontologie 4, 93-104, Berlin.

Palmqvist, P. 1997: A critical re-evaluation of the evidence for the presence of hominids in Lower Pleistocene times at Venta Micena, Southern Spain. J. Human Evol. 33, 1, 83-89, London etc.

Palmqvist, P. & Arribas, A. 2001: Taphonomic decoding of the paleobiological information locked in a lower Pleistocene assemblage of large mammals. Paleobiology 27, 3, 512-530, Jacksonville.

Palmqvist, P., Arribas, A. & Martinez-Navarro, B. 1999: Ecomorphological study of large canids from the lower Pleistocene of southeastern Spain. Lethaia 32, 75-88, Oslo.

Palmqvist, P., Martinez-Navarro, B. & Arribas, A. 1996: Prey selection by terrestrial carnivores in a lower Pleistocene palaeocommunity. Paleobiology 22, 4, 514-534, Jacksonville.

Parés, J. M. & Pérez-González, A. 1998: Contexto estratigráfico y cronológico de Gran Dolina (Yacimiento de Atapuerca). In: Aguirre, E. (ed.): Atapuerca y la Evolución Humana, 51-60, Fundación Ramón Areces, Madrid.

1999: Magnetochronology and stratigraphy at Gran Dolina section, Atapuerca (Burgos, Spain). J. Human Evol. 37, 3/4, 325-342, London etc.

Pei, W.-C. 1934: On the Carnivora from Locality 1 of Choukoutien. Palaeont. Sin., Ser. C, 8, 1, 1-224, Peiping.

Peters, G. 1993: *Canis lupus* Linnaeus, 1758 – Wolf. In: Stubbe, M. & Krapp, F. (eds.): Handbuch der Säugetiere Europas, Vol. 5: Raubsäuger – Carnivora (Fissipedia), Teil II: Mustelidae 2, Viverridae, Herpestidae, Felidae,

47-106, Aula-Verl., Wiesbaden.

Pfeiffer, T. 1997: *Dama (Pseudodama) reichenaui* (Kahlke 1996) (Artiodactyla: Cervidae, Cervini) aus den Mosbach-Sanden (Wiesbaden-Biebrich). Mainzer Naturwiss. Archiv 35, 31-59, Mainz.

1998: *Capreolus suessenbornensis* Kahlke 1956 (Cervidae, Mammalia) aus den Mosbach-Sanden (Wiesbaden-Biebrich) mit einem Beitrag zur Stellung der Rehe im System pleistozäner und holozäner Cerviden. Mainzer Naturwiss. Archiv 36, 47-76, Mainz.

1999: Die Stellung von *Dama* (Cervidae, Mammalia) im System plesiometacarpaler Hirsche des Pleistozäns. Phylogenetische Rekonstruktion – Metrische Analyse. Courier Forschungsinst. Senckenberg 211, 1-218, Frankfurt a. M.

2000: Die Cerviden der altpleistozänen Fundstelle Bad Deutsch-Altenburg 4B (Niederösterreich). Beitr. Paläont. 25, 95-103, Wien.

Rabeder, G. 1972: Die Insectivoren und Chiropteren (Mammalia) aus dem Altpleistozän von Hundsheim (Niederösterreich). Ann. Naturhist. Mus. Wien 76, 375-474, Wien.

1976: Die Carnivoren (Mammalia) aus dem Altpleistozän von Deutsch-Altenburg 2. Mit Beiträgen zur Systematik einiger Musteliden und Caniden. Beitr. Paläont. Österr. 1, 5-119, Wien.

Radulescu, C. & Kovács, A. 1966: Contributii la cunoasterea faunei de mamifere fosile din Bazinul Baraolt (depresiunea Brasov). Lucr. Inst. speol. „Emil Racovita“ 5, 233-250, Bucuresti.

1968: Noi contributii la cunoasterea faunei de mamifere fosile din Bazinul Baraolt (Depresiunea Barsov). Lucr. Inst. speol. „Emil Racovita“ 7, 231-253, Bucuresti.

Radulesco, C. & Samson, P. 1962: Sur la Presence de *Megalovis latifrons* dans le Villafranchien d'Oltenie (Roumanie). Vertebrata Palasiatica 6, 3, 262-269, Beijing (in Chinese, French summary).

1965: *Soergelia elisabethae* Schaub dans le Pléistocène moyen de l'Olténie (Roumanie). Eclogae geol. Helv. 58, 2, 1107-1110, Bâle.

1985: Pliocene and Pleistocene mammalian biostratigraphy in southeastern Transylvania (Romania). Trav. Inst. Spéol «Emile Racovitza» 24, 85-95, Bucarest.

1990: The Plio-Pleistocene Mammalian Succession of the Oltet Valley, Dacic Basin, Romania. Quartärpaläontologie 8, 225-232, Berlin.

1991: Traces d'activité humaine a la limite Pliocène / Pléistocène dans le Bassin Dacique (Roumanie). In: Bonifay, E. u. Vandermeersch, B. (dir.): Les Premiers Européens, 114è Congr. nat. Soc. sav. (Paris, 3-9 avril 1989), 203-207, Paris.

Radulesco, C., Samson, P., Mihaila, N. & Kovács, A. 1965: Contributions à la connaissance des faunes de Mammifères pléistocènes de la Dépression de Brasov (Roumanie). Eiszeitalter u. Gegenwart 16, 132-188, Öhringen.

Radulesco, C., Samson, P.-M. & Ştiucă, E. 1998: Cadre biostratigraphique du Paléolithique inférieur en Roumanie. Quaternaire 9, 4, 283-290, Paris.

Rai, U. 1998: Major Indian Reserve Devastated by Monsoon Floods. CAT News 29, 22-23, Bougy.

Rakovec, I. 1954: *Libralces* aff. *gallicus* Azzaroli z viskega Brda pri Ljubljani. *Libralces* aff. *gallicus* Azzaroli from the Vic terrace near Ljubljana. Razprave Slov. Akad. Znanosti Umetnosti, cl. 4, 2, 297-317, Ljubljana.

1975: Razvoj kvartarne sesalske favne Slovenije. Über quartäre Säugetierfaunen Sloweniens (NW Jugoslawien). Arheol. Vestnik 24, 225-270, Ljubljana.

Raynal, J.-P., Magoga, L. & Bindon, P. 1995: Tephrofacts and the first human occupation of the French Massif Central. In: Roebroeks, W. & Kolfschoten, T. v. (eds.): The Earliest Occupation of Europe, Proceedings of the European Science Foundation Workshop at Tautavel (France), 1993, 129-146, Univ. Leiden, Leiden.

Rekovec, L. I. 1994: Melkie mlekopitaûšie antropogena ûga vostočnoj Evropy. 370 p., Naukova dumka, Kiev (in Russian).

Richmond, G. M. 1996: The INQUA-approved provisional Lower-Middle Pleistocene boundary. In: Turner, C. (ed.): The early Middle Pleistocene in Europe, Proceedings of the SEQS Cromer Symposium, Norwich / United Kingdom, 3-7 September 1990, 319-327, Balkema, Rotterdam, Brookfield.

Roberts, M. B., Gamble, C. S. & Bridgland, D. R. 1995: The earliest occupation of Europe: the British Isles. In: Roebroeks, W. & Kolfschoten, T. v. (eds.): The Earliest Occupation of Europe, Proceedings of the European Science Foundation Workshop at Tautavel (France), 1993, 165-191, Univ. Leiden, Leiden.

Roberts, M. B., Stringer, C. B. & Parfitt, S. A. 1994: A hominid tibia from Middle Pleistocene sediments at Boxgrove, UK. Nature 369, 311-313, London.

Roebroeks, W. & Kolfschoten, T. v. 1995: The earliest occupation of Europe: a reappraisal of artefactual and chronological evidence. In: Roebroeks, W. & Kolfschoten, T. v. (eds.): The Earliest Occupation of Europe, Proceedings of the European Science Foundation Workshop at Tautavel (France), 1993, 297-315, Univ. Leiden, Leiden.

Rook, L. & Torre, D. 1996: The latest Villafranchian – early Galerian small dogs of the Mediterranean area. In: Nadachowski, A. & Werdelin, L. (eds.): Neogene and Quaternary mammals of the Palaearctic, Papers in mammal palaeontology honoring Kazimierz Kowalski, Acta zool. cracov. 99, 1, 427-434, Kraków.

Rosas, A., Huguet, R., Pérez-González, A., Carbonell, E., Vallverdú, J., Made, J. v. d., Allué, E., Garcia, N., Martinez-Pérez, R., Rodriguez, J., Sala, R., Saladie, P., Simón, G., Martinez-Maza, C., Bastir, M., Sánchez, A. & Parés, J. M. 2004: Initial approach to the site formation and Paleoecology of the "Sima del Elefante": a Pleistocene karst locality at Atapuerca Hill. In: Baquedano, E. & Rubio Jara, S. (eds.): Zona Arqueológica, Miscelánea en homenaje a Emiliano Aguirre, Volumen I, Geologia, 135-155, Alcalá de Henares, Museo Arqueológico Regional, Madrid.

Roth, H. H. 1969: Über das Schwimmen afrikanischer Landsäugetiere im Kariba-Stauseegebiet und ihr Verhalten gegenüber Flutwasser. Zool. Garten 37, 1/3, 12-29, Leipzig.

Roth, H. H. & Child, G. 1968: Distribution and Population Structure of Black Rhinoceros (*Diceros bicornis* L.) in the Lake Kariba basin. Z. Säugetierkd. 33, 214-226, Hamburg, Berlin.

Rutte, E. 1958: Die Fundstelle altpleistozäner Säugetiere von Randersacker bei Würzburg. Geol. Jb. 73, 737-754, Hannover.

1959: Neue Beobachtungen in der Fundstelle altpleistozäner Säugetiere von Randersacker bei Würzburg. Fränkische Natur u. Landschaft 2, 30-35, Würzburg.

1997: Die Fischreste aus dem Unterpleistozän von Untermaßfeld. In: Kahlke, R.-D. (ed.): Das Pleistozän von Untermaßfeld bei Meiningen (Thüringen), Part 1, Monogr. Röm.-Germ. Zentralmus. Mainz 40, 1, 77-80 + pl. 20, Habelt, Bonn.

Sablin, M. V. 1990: Ostatki hiŝnyh i kopytnyh iz nizneapŝeronskih otlozenij Azerbajdzana. Akad. Nauk SSSR, Trudy Zool. Inst. 213, 138-142, Leningrad (in Russian, English summary).

Sala, B. 1987: *Bison schoetensacki* Freud. from Isernia la Pineta (early Mid-Pleistocene – Italy) and revision of the european species of bison. Palaeontograph. Ital. 74, 113-170, Pisa.

1990: *Panthera leo fossilis* (v. Reich., 1906) (Felidae) de Isernia la Pineta (Pléistocène moyen inférieur d'Italie). Géobios 23, 2, 189-194, Lyon.

Sala, B., Masini, F., Ficcarelli, G., Rook, L. & Torre, D. 1992: Mammal dispersal events in the Middle and Late Pleistocene of Italy and Western Europe. In: Koenigswald, W. v. & Werdelin, L. (eds.): Mammalian Migration and Dispersal Events in the European Quaternary, Courier Forschungsinst. Senckenberg 153, 59-68, Frankfurt a. M.

Samson, P. 1976: Les Equides fossiles de Roumanie (Pliocène moyen - Pléistocène supérieur). Geol. Rom. 14, 165-352, Roma.

Samson, P. & Radulesco, C. 1965: Die Säugetierfaunen und die Grenzen Pliozän/Pleistozän und Unterpleistozän/Mittelpleistozän in Rumänien. Ber. Geol. Ges. DDR 10, 1, 67-76, Berlin.

1966: Sur la présence des Girafidés dans le Villafranchien supérieur de Roumanie. N. Jb. Geol. Paläont. Mh. 10, 588-594, Stuttgart.

1973: Les feauns [*sic!*] de Mammifères et la limite Pliocène-Pléistocène en Roumanie. Trav. Inst. Spéol «Emile Racovitza» 12, 191-228, Bucarest.

Samson, P., Radulescu, C. & Kovács, A. 1969: Faunele de mamifere si stratigrafia cuaternarului in Depresiunea Brasov. Aluta 1, 193-209, Sf. Gheorghe.

Sardella, R. 1998: The Plio-Pleistocene Old World dirk-toothed cat *Megantereon* ex gr. *cultridens* (Mammalia, Felidae, Machairodontinae), with comments on taxonomy, origin and evolution. N. Jb. Geol. Paläont. Abh. 207, 1, 1-36, Stuttgart.

Sardella, R., Caloi, L., Stefano, G. di, Palombo, M. R., Petronio, C., Abbazzi, L., Azzaroli, A., Ficcarelli, G., Mazza, P., Mezzabotta, C., Rook, L., Torre, D., Argenti, P., Capasso Barbato, L., Kotsakis, T., Gliozzi, E., Masini, F. & Sala, B. 1998: Mammal Faunal Turnover in Italy from the Middle Pliocene to the Holocene. In: Kolfschoten, T. v. & Gibbard, P. L. (eds.): The Dawn of the Quaternary, Proceedings of the SEQS-EuroMam symposium 1996, Meded. Nederl. Inst. Toegepaste Geowetenschappen TNO 60, 499-511, Haarlem.

Sardella, R., Stefano, G. di & Petronio, C. 1995: The Villafranchian mammal faunas from the Tiber river basin (Umbria, Central Italy). Il Quaternario 8, 2, 509-514, Verona.

Saunders, J. J. & Dawson, B. K. 1998: Bone damage patterns produced by extinct Hyena, *Pachycrocuta brevirostris* (Mammalia: Carnivora), at the Haro River Quarry, Northwestern Pakistan. In: Tomida, Y., Flynn, L. J. & Jacobs, L. L. (eds.): Advances in Vertebrate Paleontology and Geochronology, Nat. Sci. Mus. Monogr. 14, 215-242, Tokyo.

Schäfer, W. 1976: Aktuopaläontologische Beobachtungen. 10. Zur Fossilisation von Vögeln. Natur u. Museum 106, 9, 276-279, Frankfurt a. M.

Schmincke, H.-U. & Bogaard, P. v. d. 1995: Die Datierung des Masavera-Basaltlavastroms. Jb. Röm.-Germ. Zentralmus. Mainz 42, 51-52, Habelt, Bonn.

Schreuder, A. 1929: *Conodontes (Trogontherium)* and *Castor* from the Teglian Clay compared with the Castoridae from other localities. Arch. Mus. Teyler 3, 6, 99-318 + 16 pl., Haarlem.

Schreve, D., Currant, A. & Stringer, C. 1999: Conclusion: correlation of the Westbury Cave deposits. In: Andrews, P., Cook, J., Currant, A. & Stringer, C. (eds.): Westbury Cave. The Natural History Museum Excavations 1976-1984, 275-284, Western Academic & Specialist Press, Bristol.

Schütt, G. 1974: Die Carnivoren von Würzburg-Schalksberg. Mit einem Beitrag zur biostratigraphischen und zoogeographischen Stellung der altpleistozänen Wirbeltierfaunen vom Mittelmain (Unterfranken). N. Jb. Geol. Paläont. Abh. 147, 1, 61-90, Stuttgart.

Seidel, G., Kästner, H. & Wiefel, H. 1998: Geologische Übersichtskarte 1 : 200.000. CC 5526 Erfurt. Bundesanst. f. Geowiss. u. Rohstoffe, Hannover.

Shackleton, N. J. 1995: New Data on the Evolution of Pliocene Climatic Variability. In: Vrba, E. S., Denton, G. H., Partridge, T. C. & Burckle, L. H. (eds.): Paleoclimate and Evolution, with Emphasis on Human Origins, 242-248, Yale Univ. Press, New Haven, London.

Sher, A. V. 1986: Pâtyj paleontologičeskij kollokvium v Vejmare (GDR). Paleont. Žurn. 3, 139-141, Moskva (in Russian).

1997: An Early Quaternary Bison population from Untermaßfeld: *Bison menneri* sp. nov. In: Kahlke, R.-D. (ed.): Das Pleistozän von Untermaßfeld bei Meiningen (Thüringen), Part 1, Monogr. Röm.-Germ. Zentralmus. Mainz 40, 1, 101-180 + pls. 22-33, Habelt, Bonn.

1999: The identity of the ‚Tamanian elephant‘. In: Reumer, J. W. F. & Vos, J. de (eds.): Official Conference Papers, Volume of Abstracts, Excursion Guide, 2nd International Mammoth Conference, 200 Years of Mammoth Research, May 16 - 20, 1999, Natuurmuseum Rotterdam, The Netherlands, 56-58, Natuurmuseum Rotterdam, Rotterdam.

Sickenberg, O. 1976: Eine Säugetierfauna des tieferen Bihariums aus dem Becken von Megalopolis (Peloponnes, Griechenland). Ann. Géol. Pays Hellén. 27, 25-73, Athènes.

Sickenberg, O., unter Mitarbeit von Becker-Platen, J. D., Benda, L., Berg, D., Engesser, B., Gaziry, W., Heissig, K., Hünermann, K. A., Sondaar, P. Y., Schmidt-Kittler, N., Staesche, K., Staesche, U., Steffens, P. & Tobien, H. 1975: Die Gliederung des höheren Jungtertiärs und Altquartärs in der Türkei nach Vertebraten und ihre Bedeutung für die internationale Neogen-Stratigraphie (Känozoikum und Braunkohlen der Türkei. 17.). Geol. Jb., B, 15, 1-167, Hannover.

Sickenberg, O. & Tobien, H. 1971: New Neogene and Lower Quaternary vertebrate faunas in Turkey. Newsl. Stratigr. 1, 3, 51-61, Leiden.

Smirnov, V. S. & Korytin, N. S. 1985: Vozrastnaâ struktura i sootnošenie polov. In: Bibikov, D. I. (eds.): Volk. Proishoždenie, sistematika, morfologiâ, ekologiâ, 389-408, Nauka, Moskva (in Russian).

Sondaar, P. Y. 1986: The Island Sweepstakes. Nat. Hist. 95, 9, 50-57, New York.

Sotnikova, M. V. 1976: Verhnepliocenovye Carnivora Central'noj Azii. Izv. Akad. nauk SSSR, ser. geol. 11, 133-137, Moskva (in Russian).

1978: Novye dannye o hiŝnyh mlekopitaûših verhnego pliocena i nižnego plejstocena na territorii severo-vostoka SSSR. Bûll. Kom. Izuč. Četv. Per. 48, 22-30, Moskva (in Russian).

1988: Associaciâ melkij volk – ksenocion v pozdnem èoplejstocene - rannem plejstocene Zentral'noj Azii. Bûll. Kom. Izuč. Četv. Per. 57, 78-89, Moskva (in Russian).

1989: Hiŝnye mlekopitaûŝie pliocena - rannego plejstocena. Stratigrafičeskoe značenie. 123 p., Nauka, Moskva (in Russian).

2001: Remains of Canidae from the Lower Pleistocene site of Untermaßfeld. In: Kahlke, R.-D. (ed.): Das Pleistozän von Untermaßfeld bei Meiningen (Thüringen), Part 2, Monogr. Röm.-Germ. Zentralmus. Mainz 40, 2, 607-632 + pls. 102-113, Habelt, Bonn.

Sotnikova, M. V., Dodonov, A. E. & Pen'kov, A. V. 1997: Upper Cenozoic bio-magnetic stratigraphy of Central Asian mammalian localities. Palaeogeogr., Palaeoclim., Palaeoecol. 133, 243-258, Amsterdam.

Sotnikova, M. V. & Sablin, M. V. 1993: Pozdnevillafrankskaâ associaciâ hiŝnyh mlekopitaûših iz mestonahoždeniâ Palan-Tûkan (Vostočnoj Zakavkaz'e, Respublika Azerbajdžan). In: Baryšnikov, G. F. & Kuz'mina, I. E. (eds.): Materialy no mezozojskoj i kajnozojskoj istorii nazemnyh pozvonočnyh, Ross. Akad. Nauk, Trudy Zool. Inst. 249, 134-145, Sankt-Peterburg (in Russian, English summary).

Sotnikova, M. V. & Vislobokova, I. A. 1990: Pleistocene Mammals from Lakhuti, Southern Tajikistan, U.S.S.R. Quartärpaläontologie 8, 237-244, Berlin.

Soto, E. 1987: Grandes herbívoros del Pleistoceno medio de la Trinchera del Ferrocarril de Atapuerca (Burgos, España). In: Aguirre, E., Carbonell, E. & Bermudez de Castro, J. M. (eds.): El hombre fósil de Ibeas y el Pleistoceno de la Sierra de Atapuerca, 92-116, Junta de Castilla y Leon, Valladolid.

Spassov, N. 2003: The Plio-Pleistocene vertebrate fauna in South-Eastern Europe and the megafaunal migratory waves from the east to Europe. Revue Paléobiol. 22, 1, 197-229, Genève.

Spassov, N. & Crégut-Bonnoure, É. 1999: Premières données sur les Bovidae Villafranchiens de Bulgarie. C. R. Acad. Sci. Paris, Sci. de la terre et des planètes, 328, 493-498, Paris.

Steensma, K. J. 1988: Plio-/Pleistozäne Großsäugetiere (Mammalia) aus dem Becken von Kastoria/Grevena, südlich von Neapolis – NW-Griechenland. PhD Thesis, 315 p., Technische Universität Clausthal, Clausthal.

Steininger, F. F. & Piller, W. E. (eds.), unter Mitarbeit der Stratigraphischen Kommission in Deutschland, unter Mitarbeit der Stratigraphischen Kommission in Österreich 1999: Empfehlungen (Richtlinien) zur Handhabung der stratigraphischen Nomenklatur. Courier Forschungsinst. Senckenberg 209, 1-19, Frankfurt a. M.

Stephan, E., Uerpmann. H.-P. & Cramer, B. 2001: Paläotemperaturbestimmungen nach $\delta^{18}O$-Werten von Equiden-Resten aus dem Unterpleistozän von Untermaßfeld. In: Kahlke, R.-D. (ed.): Das Pleistozän von Untermaßfeld bei Meiningen (Thüringen), Part 3, Monogr. Röm.-Germ. Zentralmus. Mainz 40, 3, 917-930, Habelt, Bonn.

Stuart, A. J. 1974: Pleistocene history of the British vertebrate fauna. Biol. Rev. 49, 225-266, London.

1979: Pleistocene occurrences of the European pond tortoise (*Emys orbicularis* L.) in Britain. Boreas 8, 359-371, Oslo.

1980: The vertebrate fauna from the interglacial deposits at Sugworth, near Oxford. Phil. Trans. R. Soc. London, B 289, 87-97, London.

1981: A Comparison of the Middle Pleistocene Mammal Faunas of Voigtstedt (Thuringia, GDR) and West Runton (Norfolk, England). Quartärpaläontologie 4, 155-163, Berlin.

1988: Preglacial Pleistocene vertebrate faunas of East Anglia. In: Gibbard, P. L. & Zalasiewicz, J. A. (eds.): Pliocene – Middle Pleistocene of

East Anglia, Field Guide, 57-64, Quaternary Research Association, Cambridge.

1992: The Pleistocene Vertebrate Faunas of West Runton, Norfolk, England. Cranium 9, 2, 77-84, Utrecht.

1996: Vertebrate faunas from the early Middle Pleistocene of East Anglia. In: Turner, C. (ed.): The early Middle Pleistocene in Europe, Proceedings of the SEQS Cromer Symposium, Norwich / United Kingdom, 3-7 September 1990, 9-24, Balkema, Rotterdam, Brookfield.

1997: The West Runton elephant discovery and excavation. 12 p., Norfolk Museums Service, Norwich.

Symeonidis, N. & Vos, J. de 1977: Grosssäuger-Funde aus den plistozänen Spaltenfüllungen von Turkovunia in Athen. Ann. Géol. Pays Hellén. 28, 135-144, Athènes.

Symeonidis, N. & Zapfe, H. 1977: Primatenzähne (Cercopithecidae) aus einer plistozänen Spaltenfüllung im Steinbruch Tourkobounja, Athen. Ann. Géol. Pays Hellén. 28, 207-214, Athènes.

Tappen, M., Adler, D. S., Ferring, C. R., Gabunia, M., Vekua, A. & Swisher III, C. C. 2002: Akhalkalaki: The Taphonomy of An Early Pleistocene Locality in the Republic of Georgia. J. Archaeol. Sci. 29, 1367-1391, London etc.

Tchernov, E. 1987: The Age of the 'Ubeidiya Formation, an Early Pleistocene Hominid Site in the Jordan Valley, Israel. Isr. J. Earth Sci. 36, 3-30, Jerusalem.

1988: The biogeographical history of the southern Levant. In: Yom-Tov, Y. & Tchernov, E. (eds.): The zoogeography of Israel, 159-250, Junk, Dordrecht.

1999: The Earliest Hominids in the Southern Levant. In: Gibert, J., Sánchez, F., Gibert, L. & Ribot, F. (eds.): Los Hominidos y su entorno en el Pleistoceno inferior y medio de Eurasia, Actas del Congreso Internacional de Paleontologia Humana, Orce 1995, 389-406, Museo Prehistoria y Paleontologia «J. Gibert», Orce.

Tchernov, E. & Guérin, C. (eds.) 1986: Les mammifères du Pléistocène inférieur de la vallée du Jourdain à Oubéidiyeh. Mém. trav. Centre Rech. franç. Jérusalem 5, 1-405, Paris.

Tchernov, E., Horwitz, L. K., Ronen, A. & Lister, A. 1994: The Faunal Remains from Evron Quarry in Relation to Other Lower Paleolithic Hominid Sites in the Southern Levant. Quaternary Research 42, 3, 328-339, Washington.

Terzea, E. 1993: Évènements évolutifs et de dispersion chez les rongeurs au Quaternaire de Roumanie. Trav. Inst. Spéol. «Emile Racovitza» 32, 107-118, Bucarest.

1995: Mammalian events in the Quaternary of Romania and correlations with the climatic chronology of Western Europe. Acta zool. cracov. 38, 1, 109-120, Kraków.

1996: Biochronology of the Pleistocene deposits at Betfia (Bihor, Romania). In: Nadachowski, A. & Werdelin, L. (eds.): Neogene and Quaternary Mammals of the Palaearctic, Papers in Mammal Palaeontology honoring Kazimierz Kowalski, Acta zool. cracov. 39, 1, 531-540, Kraków.

Terzea, E. & Jurcsák, T. 1968: Bemerkungen über die mittelpleistozänen Faunen von Betfia. Ber. deutsch. Ges. geol. Wiss., A, Geol. Paläont., 13, 3, 381-390, Berlin.

1969: Contributii la cunoasterea faunelor pleistocene medii de la Betfia (România). Trav. Inst. Spéol "Emile Racovitza" 8, 201-213, Bucarest.

1976: Faune de mammifères de Betfia-XIII (Bihor, Roumanie) et son âge géologique. Trav. Inst. Spéol «Emile Racovitza» 15, 195-205, Bucarest.

Thenius, E. 1954: Gepardreste aus dem Altquartär von Hundsheim in Niederösterreich. N. Jb. Geol. Paläont. Mh., 225-238, Stuttgart.

Thomas, H., Geraads, D., Janjou, D., Vaslet, D., Memesh, A., Billiou, D., Bocherens, H., Dobigny, G., Eisenmann, V., Gayet, M., Lapparent de Broin, F. de, Petter, G. & Halawani, M. 1998: First Pleistocene faunas from the Arabian Peninsula: An Nafud desert, Saudi Arabia. C. R. Acad. Sci. Paris, Sci. de la terre et des planètes, 326, 145-152, Paris.

Tiedemann, R., Sarnthein, M. & Shackleton, N. J. 1994: Astronomic timescale for the Pliocene Atlantic $\delta^{18}O$ and dust flux records of Ocean Drilling Program site 659. Paleoceanography 9, 4, 619-638, Washington.

Titov, V. V. 1999: Fauna krupnyh mlekopitaûših pozdnego pliocena severo-vostočnogo Priazov'â. Avtoreferat dissertazii, 24 p., Paleont. inst. Ross. Akad. nauk, Moskva (in Russian).

Torre, D., Albianelli, A., Azzaroli, A., Ficcarelli, G., Magi, M., Napoleone, G. & Sagri, M. 1993: Palaeomagnetic calibration of Late Villafranchian mammalian faunas from the Upper Valdarno, Central Italy. Mem. Soc. Geol. Ital. 49, 335-344, Pisa.

Torre, D., Albianelli, A., Bertini, A., Ficcarelli, G., Masini, F. & Napoleone, G. 1996: Paleomagnetic calibration of Plio-Pleistocene mammal localities in central Italy. In: Nadachowski, A. & Werdelin, L. (eds.): Neogene and Quaternary Mammals of the Palaearctic, Papers in Mammal Palaeontology honoring Kazimierz Kowalski, Acta zool. cracov. 39, 1, 559-570, Kraków.

Torre, D., Rook, L. & Ficcarelli, G. 1999: The mammal faunal turnover in Italy at the Early to Middle Pleistocene transition. In: Gibert, J., Sánchez, F., Gibert, L. & Ribot, F. (eds.): Los Hominidos y su entorno en el Pleistoceno inferior y medio de Eurasia, Actas del Congreso Internacional de Paleontologia Humana, Orce 1995, 541-548, Museo Prehistoria y Paleontologia "J. Gibert", Orce.

Torres, T., Llamas, J. F., Canoira, L., García-Alonso, P., García-Cortés, A. & Mansilla, H. 1997: Amino acid chronology of the Lower Pleistocene deposits of Venta Micena (Orce, Granada, Andalusia, Spain). Org. Geochem. 26, 1/2, 85-97, Oxford etc.

Tournepiche, J. F. 1984: Faunes et industries antérissiennes de la grotte d'Artenac (Charente). C. R. Acad. Sci. Paris 298, Sér. II, 9, 423-428, Paris.

1985: Biochronologie des faunes antéwürmiennes de Charente. Bull. Soc. Anthrop. Sud-Ouest. 20, 2-3, 131-143, Bordeaux.

1993: Rapport sur les fouilles effectuées a Artenac (Charente). Campagne 1993. unpubl., 126 p., Musée d'Angoulême, Angoulême.

1994: Rapport sur les fouilles effectuées a Artenac (Charente). Campagne 1994. unpubl., 64 p., Musée d'Angoulême, Angoulême.

Tsoukala, E. 1989: Contribution to the study of the Pleistocene fauna of large mammals (Carnivora, Perissodactyla, Artiodactyla) from Petralona Cave. Chalkidiki (N. Greece). PhD Thesis, Aristotle Univ. Thessaloniki, School of Geology, Sci. Annals 1, 8, 1-360 + pls. I-LXII, Thessaloniki (in Greek, English summary).

1991: Contribution to the study of the Pleistocene fauna of large mammals (Carnivora, Perissodactyla, Artiodactyla) from Petralona Cave (Chalkidiki, N. Greece). Preliminary report. C. R. Acad. Sci. Paris 312, Sér. II, 331-336, Paris.

1992: Quaternary faunas of Greece. In: Koenigswald, W. v. & Werdelin, L. (eds.): Mammalian Migration and Dispersal Events in the European Quaternary, Courier Forschungsinst. Senckenberg 153, 79-92, Frankfurt a. M.

1996: Comparative study of ursid remains from the Quaternary of Greece, Turkey and Israel. In: Nadachowski, A. & Werdelin, L. (eds.): Neogene and Quaternary Mammals of the Palaearctic, Papers in Mammal Palaeontology honoring Kazimierz Kowalski, Acta zool. cracov. 39, 1, 571-576, Kraków.

Turner, A. 1992a: Large carnivores and earliest European hominids: changing determinants of resource availability during the Lower and Middle Pleistocene. J. Human Evol. 22, 109-126, London etc.

1992b: Villafranchian-Galerian larger carnivores of Europe: dispersions and extinctions. In: Koenigswald, W. v. & Werdelin, L. (eds.): Mammalian Migration and Dispersal Events in the European Quaternary, Courier Forschungsinst. Senckenberg 153, 153-160, Frankfurt a. M.

1994: Evolution and Dispersion of Larger Mammals in Europe During the Time Span of *Homo erectus*. In: Franzen, J. L. (eds.): 100 Years of Pithecanthropus, The *Homo erectus* Problem, Courier Forschungsinst. Senckenberg 171, 241-247, Frankfurt a. M.

1995: The Villafranchian large carnivore guild: geographic distribution and structural evolution. Il Quaternario 8, 2, 349-356, Verona.

1999: Larger carnivores (Mammalia, Carnivora) from Westbury Cave. In: Andrews, P., Cook, J., Currant, A. & Stringer, C. (eds.): Westbury Cave. The Natural History Museum Excavations 1976-1984, 175-193, Western Academic & Specialist Press, Bristol.

2001: Remains of *Pachycrocuta brevirostris* (Carnivora, Hyaenidae) from the Lower Pleistocene site of Untermaßfeld. In: Kahlke, R.-D. (ed.): Das Pleistozän von Untermaßfeld bei Meiningen (Thüringen), Part 2, Monogr. Röm.-Germ. Zentralmus. Mainz 40, 2, 673-690 + pls. 121-127, Habelt, Bonn.

Turner, A. & Antón, M. 1996: The Giant Hyaena, *Pachycrocuta brevirostris* (Mammalia, Carnivora, Hyaenidae). Géobios 29, 4, 455-468, Villeurbanne.

1997: The Big Cats and Their Fossil Relatives. An Illustrated Guide to Their Evolution and Natural History. 234 p., Columbia Univ. Press, New York.

Turner, C. 1975: Der Einfluß großer Mammalier auf die interglaziale Vegetation. Quartärpaläontologie 1, 13-19, Berlin.

Turner, E. 1990: Middle and Late Pleistocene Macrofaunas of the Neuwied Basin Region (Rhineland-Palatinate) of West Germany. Jb. Röm.-Germ. Zentralmus. Mainz 37, 135-403 + pls. 15-32, Habelt, Bonn.

Tû'tkova, L. A. 1988: Novye dannye po pozdnepliocenovoj faune melkih mlekopitaûŝih uročiša Aktogaj. In: Mezregional'noe sopostavlenie mezozoj-kajnozojskih faun i flor Kazahstana, Materialy ist. fauny flory Kazahstana 10, 82-93, Nauka, Alma-Ata (in Russian).

Unger, K. P. & Kahlke, R.-D. 1995: Thüringen. In: Benda, L. (ed.): Das Quartär Deutschlands, 199-219, Borntraeger, Berlin, Stuttgart.

Valoch, K. 1995: The earliest occupation of Europe: Eastern Central and Southeastern Europe. In: Roebroeks, W. & Kolfschoten, T. v. (eds.): The Earliest Occupation of Europe, Proceedings of the European Science Foundation Workshop at Tautavel (France), 1993, 67-80, Univ. Leiden, Leiden.

Vangengejm, E. A., Pevzner, M. A. & Tesakov, A. S. 1990: Magnito- i biostratigrafičeskie issledovaniâ v stratoregione psekupskogo faunističeskogo kompleksa mlekopitaûŝih. Bûll. Kom. Izuč. Četv. Per. 59, 81-93, Moskva (in Russian).

Vangengejm, E. A., Vekua, M. L., Zegallo, V. I., Pevzner, M. A., Taktakisvili, I. G. & Tesakov, A. S. 1991: Položenie tamanskogo faunističeskogo komplexa v stratigrafičeskoj i magnitohronologičeskoj skalah. Bûll. Kom. Izuč. Četv. Per. 60, 41-52, Moskva (in Russian).

Vekua, A. K. 1959: O gippopotame iz nizneplejstocenovyh otlozenij Gruzij. Soob. Akad. Nauk Gruzinskoj SSR 13, 5, 561-566, Tbilisi (in Russian).

1962: Ahalkalakskaâ nižneplejstocenovaâ fauna mlekopitaûŝih. 207 p. + XIX pls., Izd. Akad. Nauk Gruzinskoj SSR, Tbilisi (in Georgian, Russian and English summary).

1986: The Lower Pleistocene Mammalian Fauna of Akhalkalaki (Southern Georgia, USSR). Palaeontograph. Ital. 74, 63-96, Pisa.

1991: Istoriâ životnogo mira – pozvonočnye. In: Gruziâ v antropogene, 308-381, Sakartvelo, Tbilisi (in Russian).

1995: Die Wirbeltierfauna des Villafranchium von Dmanisi und ihre biostratigraphische Bedeutung. Jb. Röm.-Germ. Zentralmus. Mainz 42, 77-180 + pls. 7-54, Habelt, Bonn.

Vekua, A. K., Dzigauri, D. G. & Torozov, R. I. 1985: Novye paleontologičeskie nahodki v okrestnostâh Zalki. Bull. Acad. Sci. Georgian SSR 118, 2, 373-376, Tbilisi (in Russian, English summary).

Vekua, A. & Gabunia, L. 1989: Die Wirbeltiere von Dmanisi und ihre stratigraphische Stellung. In: Dzaparidze, V., Bosinski, G., Bugianisvili, T., Gabunia, L., Justus, A., Klopotovskaja, N., Kvavadse E., Lordkipanidze, D., Majsuradze, G., Mgeladze, N., Nioradze, M., Pavenisvili, E., Schmincke, H.-U., Sologasvili, D., Tusabramisvili, D., Tvalcrelidze, M. & Vekua, A.: Der altpaläolithische Fundplatz Dmanisi in Georgien (Kaukasus), Jb. Röm.-Germ. Zentralmus. Mainz 36, 83-91 + pls. 3-5, Habelt, Bonn.

Vekua, A. & Lordkipanidze, D. 1998: The Pleistocene paleoenvironment of the Transcaucasus. Quaternaire 9, 4, 261-266, Paris.

Vekua, A., Lordkipanidze, D., Rightmire, G. P., Agusti, J., Ferring, R., Maisuradze, G., Mouskhelishvili, A., Nioradze, M., Leon, M. P. de, Tappen, M., Tvalchrelidze, M. & Zollikofer, C. 2002: A New Skull of Early *Homo* from Dmanisi, Georgia. Science 297, 85-89, Washington.

Verosub, K. L., Goren-Inbar, N., Feibel, C. S. & Saragusti, I. 1998: Location of the Matuyama/ Brunhes boundary in the Gesher Benot Ya'aqov archaeological site. Israel. J. Human Evol. 34, A22, London etc.

Vervoort-Kerkhoff, Y. & Kolfschoten T. v. 1987: Tegelen. Cranium 4, 1, 20-23, Dieren.

Vislobokova, I. A., Sotnikova, M. V. & Erbaeva, M. A. 1995: The Villafranchian mammalian faunas of the asiatic part of former USSR. Il Quaternario 8, 2, 367-376, Verona.

Vos, J. de, Mol, D. & Reumer, J. W. F. 1995: Early Pleistocene Cervidae (Mammalia, Artiodactyla) from the Oosterschelde (the Netherlands), with a revision of the cervid genus *Eucladoceros* Falconer, 1868. Deinsea 2, 95-121, Rotterdam.

1999: Fossils from the North Sea and dredging pits in the Netherlands. In: Reumer, J. W. F. & Vos, J. de (eds.): Official Conference Papers, Volume of Abstracts, Excursion Guide, 2nd International Mammoth Conference, 200 Years of Mammoth Research, May 16 - 20, 1999, Natuurmuseum Rotterdam, The Netherlands, 100-104, Natuurmuseum Rotterdam, Rotterdam.

Walter, H. & Breckle, S.-W. 1986: Ökologie der Erde. Vol. 3: Spezielle Ökologie der Gemäßigten und Arktischen Zonen Euro-Nordasiens. Zonobiom VI - IX. IX + 587 p., Fischer, Stuttgart.

Weigelt, J. 1927: Rezente Wirbeltierleichen und ihre paläobiologische Bedeutung. 227 p., Weg, Leipzig.

Wiegank, F. 1997: Paläomagnetische Charakteristik des Unterpleistozäns von Untermaßfeld. In: Kahlke, R.-D. (ed.): Das Pleistozän von Untermaßfeld bei Meiningen (Thüringen), Part 1, Monogr. Röm.-Germ. Zentralmus. Mainz 40, 1, 63-69, Habelt, Bonn.

Wolsan, M. 1990: Lower Pleistocene Carnivores of Poland. Quartärpaläontologie 8, 277-280, Berlin.

1993: Évolution des Carnivores quaternaires en Europe centrale dans leur contexte stratigraphique et paléoclimatique. L'Anthropologie 97, 2/3, 203-222, Paris.

2001: Remains of *Meles hollitzeri* (Carnivora, Mustelidae) from the Lower Pleistocene site of Untermaßfeld. In: Kahlke, R.-D. (ed.): Das Pleistozän von Untermaßfeld bei Meiningen (Thüringen), Part 2, Monogr. Röm.-Germ. Zentralmus. Mainz 40, 2, 659-671, Habelt, Bonn.

Zagwijn, W. H. & Jong, J. de 1984: Die Interglaziale von Bavel und Leerdam und ihre stratigraphische Stellung im niederländischen Früh-Pleistozän. Meded. Rijks Geol. Dienst 37, 3, S. J. Dijkstra Memorial Vol., 155-169, Haarlem.

Zapfe, H. 2001: Zähne von *Macaca* aus dem Unterpleistozän von Untermaßfeld. In: Kahlke, R.-D. (ed.): Das Pleistozän von Untermaßfeld bei Meiningen (Thüringen), Part 3, Monogr. Röm.-Germ. Zentralmus. Mainz 40, 3, 889-895 + pl. 146, Habelt, Bonn.

Zeissler, H. 1981: Schnecken und Muscheln in und um Weimar. Eine Molluskenfauna des Gebietes Weimar. Weimarer Schr. Heimatgesch. Naturkd. 44, 1-111, Weimar.

Zylkibaev, K. Z. 1975: Drevnie slony Kazahstana. 130 p., Nauka, Alma-Ata (in Russian).

Legend

with: R.-D. Kahlke,

Untermassfeld – A late Early Pleistocene (Epivillafranchian) fossil site near Meiningen (Thuringia, Germany) and its position in the development of the European mammal fauna. BAR International Series

Incompletely documented finds (cf. Section 4.1. and footnote 3)

Depths of the finds related to the datum level of the site (=1.0 m above the main plane):

< 1.5-1.0 m above datum level

< 1.0-0.5 m above datum level

< 0.5-0.0 m above datum level

> 0.0-0.5 m below datum level

> 0.5-1.0 m below datum level

> 1.0 m below datum level

Large boulder (from a key bed of the Lower Muschelkalk)

325 Number of the plan square (1 m^2 each)

Outer limit of the excavated area (situation in December 1995)

www.ingramcontent.com/pod-product-compliance
Lightning Source LLC
LaVergne TN
LVHW072257300926
849160LV00018B/1688